CW00375621

MUK003220

A PRACTICAL APPROACH TO HAZARD IDENTIFICATION
For Operations and Maintenance Workers

A PRACTICAL APPROACH TO HAZARD INDENTIFICATION

For Operations and Maintenance Workers

Center for Chemical Process Safety
New York, New York

A JOHN WILEY & SONS, INC., PUBLICATION

A Joint Publication of the Center for Chemical Process Safety of the American Institute of Chemical Engineers and John Wiley & Sons, Inc.

Published by John Wiley & Sons, Inc., Hoboken, New Jersey.
Published simultaneously in Canada.

For general information on our other products and services or for technical support, please contact our Customer Care Department within the United States at (800) 762-2974, outside the United States at (317) 572-3993 or fax (317) 572-4002.

Wiley also publishes its books in a variety of electronic formats. Some content that appears in print may not be available in electronic format. For information about Wiley products, visit our web site at www.wiley.com.

Library of Congress Cataloging-in-Publication Data:

A practical approach to hazard identification for operations and maintenance workers.
    p. cm.
  Includes index.
  ISBN 978-0-470-63524-7 (cloth/cd)
  1. Hazardous substances—United States—Safety measures. 2. Chemical industry—Safety measures.
I. American Institute of Chemical Engineers. Center for Chemical Process Safety
  T55.3.H3P73 2010
  660'.2804—dc22   2010004246

Printed in the United States of America.

10 9 8 7 6 5 4 3 2 1

Contents

5

EVALUATE HAZARDS 151

6

7

8

LEARNING AND CONTINUOUS IMPROVEMENT 253

PREFACE

The ability to recognize and respond to hazards at the field level is important to the safety of workers at any facility. While hazard identification may be viewed as a first step towards formal risk analysis, some situations require immediate action if serious incidents are to be avoided. This book is about identifying, ranking and addressing workplace hazards so that all workers can return home safely at the end of a work day.

As an experienced process operator or maintenance technician, you have acquired considerable skills and knowledge. Some of this knowledge makes you aware of workplace hazards such as pinch points, moving machine parts or objects falling from elevated heights. That knowledge was likely gained with time through minor encounters resulting in the wisdom of "common sense". Process or chemical hazards, though relatively safe when they are enclosed within equipment, can result in fires, explosions or toxic effects when they escape to the open air. You may not be aware of your exposure to these during normal work. Process hazards are less obvious than physical hazards and cannot always be detected at first glance. Consequently, some additional effort and different strategies may be required to identify process hazards so that they are addressed appropriately.

This book contains many photographs of hazards, abnormal situations and consequences of incidents. It explains what was learned and how you might use this information along with your skills and senses to improve workplace safety. We hope that you are able to apply the material provided here to better anticipate and prevent problems from occurring. In the future, if you find yourself effectively managing a hazardous situation with the help of this book, then we will have succeeded in our endeavor. The photographs and figures presented in this book were either (a) taken from public sources, (b) contributed by subcommittee members, or (c) created specifically for this book. CCPS is grateful for the contribution of photographs and figures.

The Center for Chemical Process Safety (CCPS) has published over 90 reference textbooks dealing with various aspects of process safety since 1989. Many of these were written from a technical perspective to help with program implementation. This concept book is our first CCPS publication intended for field use and we hope it will provide value to all users.

ACKNOWLEDGMENTS

The American Institute of Chemical Engineers (AIChE) and the Center for Chemical Process Safety (CCPS) express their appreciation and gratitude to all members of the Hazard Identification Subcommittee and their CCPS member companies for their generous support and technical contributions in the preparation of this concept book. The AIChE and CCPS also express their gratitude to the team of authors from RRS/Schirmer, an Aon Global company.

HAZARD IDENTIFICATION SUBCOMMITTEE MEMBERS:

Robert Wasileski, Chair	NOVA Chemicals
Todd Aukerman	LANXESS Corporation
Joyce Becker	BP
Susie Cowher	ISP Technologies
Christy Franklyn	RRS/Schirmer, an Aon Global company
Tim Humbke	Shell Canada
Dan Isaacson	Lubrizol
Brian Kelly	CCPS Staff Consultant
Jim Klein	Dupont
David Lewis	Occidental Chemical
Fred Simmons	Savannah River Nuclear Solutions
Jim VanOmmeren	Air Products and Chemicals, Inc.
Della Wong	Shell Canada

CCPS wishes to acknowledge the many contributions of the RRS/Schirmer staff members who wrote this book, especially the principal authors:

Dennis Attwood

Christy Franklyn

The authors wish to thank the following RRS/Schirmer personnel for their technical contributions and review: John Alderman and Tom Lawrence. Additional support was provided by Donna Pruitt, Cassie Trebilcock and Teresa Grygiel.

Before publication, all CCPS books are subjected to a thorough peer review process. CCPS gratefully acknowledges the thoughtful comments and suggestions of the peer reviewers. Their work enhanced the accuracy and clarity of these guidelines.

Peer Reviewers:

Johnnie Banks	US Chemical Safety Board
Richard Blais	Worksafe NB (New Brunswick)
Thom Bouis	Hygeia Group LLC
Andy Hart	NOVA Chemicals
Don Lorenzo	ABS Consulting
Lisa Morrison	BP
John Shrives	Environment Canada

Additional input was provided to the committee by Paul Brimmer, Paul Cameron, Shane Fuson, Mark Futrell, Monica Garner, Raymond Gould, Geoff Smith, Rich Verwegen, and George Yoksas.

Items on the CD Accompanying this Book

Folder: Pictures

A Practical Approach to Hazard Identification Paper

CCPS A Practical Approach to Hazard Identification Presentation

CCPS Hazard Identification Logo

CCPS Hazard Identification

Contractor Job Site Audit

Facility Siting Human Factors Checklist

Haz ID Loss Event List

Hazard Awareness Training Presentation

Hazard Hunter

Hazard Presentation

MOC Risk Level Checklist

NOAA Output

Observation Card

Process Safety PGI

Reaction Matrix

Risk Based Decisions

Spot the Hazard

Task Analysis Worksheet

1

INTRODUCTION

History suggests that incidents tend to repeat themselves (Ref. 1-1). Even with tightened safety regulations, process industries continue to be plagued with incidents that result in personnel injury, community impact, environmental impact, damage to the facility, and production downtime. Most of these incidents are the result of hazards that exist in our workplace.

Many incidents are similar to those that have previously occurred in industry. These events provide us with the ability to better recognize hazards and prevent recurrence. *How can we learn to manage these hazards and reduce the risk at our facilities?*

The pictures below illustrate the results of hazards that were not identified or controlled.

In simple terms, a "hazard" is a source of harm. Hazardous conditions or situations can cause harm to people and damage to property and environment when the hazards are not controlled. The term "dangerous" is sometimes used in a similar manner, but generally describes a more imminent situation, where the likelihood of harm is greater. Dangerous situations may even require an operation to be suddenly shut down for a temporary period of time, whereas hazardous conditions may be present during normal operations.

Risk is the combination of likelihood and consequence of an event that can cause harm. Identifying hazards is crucial to managing risk at a facility. To properly identify hazards, it is important to realize the following concepts:

- Hazards are conditions that can directly cause harm to people and/or damage to the environment and equipment.
- Hazards may occur naturally, may be inherent to the nature of material, or may be the result of some poorly designed or poorly managed process or activity.
- Hazards in the workplace are associated with chemicals, process equipment, operating conditions, physical activities and the general work environment.

Hazards are all around us. The challenge is to recognize hazards *and then to do something about them!*

Hazards become more of a concern when people or property can be impacted by the hazard. A snake in the grass will only pose a risk to humans if they come close to the snake (Figure 1-1).

Figure 1-1.   Snake in the grass

Common workplace hazards and their effects on people are summarized in Table 1-1.

Table 1-1.   Common hazards and their effects on people

Hazard Types	Example	Potential Impact to People
Heat	Touching hot pipe Excessive ambient temperature (heat stress) Over-exertion Working near hot equipment	Burn Heat stress or heat stroke Fatality
Cold	Brittle fracture causing equipment failure Low ambient temperatures High pressure gas release	Frostbite Touch-freeze adhesion of skin Hypothermia Fatality
Electrical	Touching live wire/terminal Equipment not grounded	Burn Electrical shock Fatality
Impact	Motor vehicle impact Falling ice Dropped objects Projectiles	Broken bones Head injury Fatality
Noise	High pitch sound Equipment noise Excessive noise levels	Reduced hearing ability Loss of hearing
Chemical	Mixing reactive chemicals Spill	Chemical burn Exposure to vapors Respiratory damage Fatality

Table 1-1. Common hazards and their effects on people (continued)

Hazard Types	Example	Potential Impact to People
Vibration	Whole body (extended driving over unpaved road, working on a compressor deck, etc.) Extremities (use of power tool)	Muscular sprain Circulation problems Raynaud's disease
Radiation	Glare Lighting Laser exposure X-ray exposure	Burn Temporary loss of sight Permanent loss of sight Cancer Fatality
Biological	Sampling water cooling towers Drinking non-potable water Stepping on a sharp object Rotting or decaying materials Bird droppings	Sickness Disease Fatality
Rotation	Turning valves Reaching for valve Entanglement with rotating shaft Rotating equipment	Muscular sprain Dismemberment
Airborne particulates	Dust Asbestos dust Sandblasting Changing filter element	Respiratory damage Particulates in eyes
Excavation	Trench collapse Damage to underground pipeline resulting in release of process materials	Body injury Suffocation Exposure to hazardous material
Asphyxiation	Inadequate oxygen Inert entry (nitrogen) Improper confined space protocols Improper rescue of co-worker Improper connection to air supply	Loss of consciousness Fatality

This table is provided for illustrative purposes only and describes some common industry hazards and their effects on people.

Although this table focuses on occupational or personnel hazards, process hazards that cause equipment failure and release of toxic or flammable material can also have a significant impact on workers at the site and will be discussed in further detail in Chapter 5.

The motivation for identifying and managing hazards is simple:

- We want ourselves, our friends, and coworkers to go home safely every day.
- We want to keep our jobs; the economic impact of major process incidents can cause a plant or facility to shut down.
- We want to protect our community and neighbors; process incidents can have consequences for communities located around them.
- We all share the environment and irreversible events can cause long-term damage to the environment and the living things that coexist in the same spaces.

An effective hazard identification and risk assessment program should motivate workers to accept personal responsibility and manage their own safety.   Workers should be empowered to immediately stop any situations or behaviors that present hazards to people, the environment and equipment.   To achieve this objective, employees must be trained to recognize hazards throughout the facility and provide solutions to correct them.

HAZARD IDENTIFICATION IS NOT OPTIONAL – IT'S AN ESSENTIAL PART OF DAY-TO-DAY ACTIVITIES.

While a hazard identification program is not a replacement for a formal Process Hazard Analysis (PHA), it is another important component of risk management. It should be recognized that some "hidden" hazards are not easily detectable by front-line workers and others will be created by the workers themselves. Therefore, there is a strong incentive to closely examine processes and equipment both before and during work to identify hazards and reduce the risk of an incident.

The goal of an effective hazard identification program is to establish a workplace built on a sense of achievement, recognition, responsibility, group decision-making, and job enrichment.   By instilling a strong safety culture into the work environment, people will become more involved and take ownership of the hazard management objectives.

An effective hazard identification program will help prevent incidents, like the one discussed below:

IMPORTANT WARNING

A process operator in a gas plant wanted to vent gas from a 6" diameter jump-over line. Recognizing that the gas contained hydrogen sulfide, the operator suited up with a self-contained breathing apparatus. This proved to be a wise move. When the operator turned the 6" valve handle one-quarter revolution, the bonnet blew off the valve body, just missing his face. He was later able to safely isolate the line from another location.

A formal investigation revealed that wet sour gas had penetrated the packing and leaked into the cavity between the bonnet and the packing. Severe corrosion was evident on the inner bonnet and on the bolts that secured it in place. The slight torquing of the valve stem was enough to cause the bolts to fail, releasing the bonnet and packing under pressure.

It is important to regularly inspect valves in sour or corrosive service. If there is any sign of corrosion on the threads or casing, a closer examination may be warranted.

Some people would consider themselves lucky to survive such an event and then focus on fixing the failed valve. However, next time the worker, or his colleague, may not be so lucky. Any high energy release associated with the abrupt failure of a piece of equipment can cause serious injury or death. Incidents such as this should trigger a formal investigation to determine how and why the equipment failed.

How far reaching is the problem and what can be done to prevent similar incidents in the future?

Were there any external symptoms that might have been detected in the field to alert workers of the immediate hazard?

Or was this a "hidden" hazard?

This book provides guidance for identifying and controlling hazards in the workplace to help:

- Raise hazard recognition awareness
- Improve your ability to detect hazards
- Empower you to take action and follow-up
- Prevent injuries and accidents
- Pass this safety culture on to new employees

Even though industry's focus has become hazard elimination and management through inherently safer design, the nature of a process facility makes it difficult to design-out all hazards. The characteristics of materials/products that make them hazardous are often what make them valuable (e.g., the flammability of gasoline).

Inherently safer design is an approach to hazard elimination and minimization (See Chapter 6.3.1). Hazards exist in the flammable and toxic materials that are processed, in the equipment used to perform work, and in the work conditions and environment.

A process for effectively recognizing and addressing hazards is illustrated in Figure 1-2. The core of the figure describes the types of hazards associated with a process facility. The outer band illustrates the steps in identifying hazards, evaluating their risk, making risk-based decisions for risk reduction, and implementing a hazard management program to mitigate the hazards and risks. This figure will reappear throughout the book to describe how each Chapter builds part of the hazard management process. The process for addressing risk through hazard identification and elimination/mitigation includes the following steps:

- Understanding basic concepts *(Chapter 2)*
- Identifying hazards *(Chapter 3)*
- Understanding different types of hazards and their severity *(Chapter 4)*
- Evaluating hazards *(Chapter 5)*
- Making risk-based decisions *(Chapter 6)*
- Implementing a hazard management program and making it part of a facility's culture *(Chapter 7)*
- Ensuring we learn from past mistakes *(Chapter 8)*

IMPLEMENT
*Mitigate the hazards
and risks*

IDENTIFY
Identify the hazards

DECIDE
*Make risk-based
decisions*

EVALUATE
*Understand the hazards
and risks*

PROCESS PHYSICAL CHEMICAL

HAZARDS

EQUIPMENT ELECTRICAL

Figure 1-2.   Hazard management process

There are two essential elements in establishing an effective hazard management program:

- Management commitment to providing the resources, empowering personnel, and measuring and managing the process.
- Employee ownership of the program to ensure that it is effectively implemented and maintained.

We've all heard the words "Safety Takes No Vacation" or "Safety Doesn't Take a Holiday", yet incidents in industry continue to remind us that hazard identification and risk management require vigilance and continuous improvement (Figure 1-3).

Figure 1-3.   The need for prevention

Fires and explosions that occur in manufacturing facilities can have widespread, devastating consequences.

1.1   INTENDED AUDIENCE

This book is primarily intended to provide industrial plant operations and maintenance workers with practical methods for identifying and managing physical and process hazards. This book should also provide benefit for people who:

- Are planning on participating in a formal Process Hazards Analysis (PHA) or safety review
- Occasionally enter a process facility and have not received formal training
- Implement new designs or review/approve Management of Changes in an existing operating facility
- Are responsible for providing resources for hazard control and elimination
- Are safety inspectors or regulators
- Are safety professionals
- Are new to the workforce

This book will help you identify and manage workplace hazards (Figure 1-4).

Figure 1-4.   Workplace activities with potential for hazards
(See color insert)

Construction and maintenance activities can create new and unique hazards.   For example, overhead loads suspended by a crane create potential energy hazards, while entry into confined spaces and vessels may create asphyxiation hazards.

The activities pictured above are common on a construction site and represent a high level of energy. Heavy loads and moving equipment can easily distract a worker and direct his or her attention from the work at hand. Construction hazards are particularly critical when modifications are made on an operating plant site. A vehicle impact or collapsing crane boom can damage delicate process equipment, releasing the contents to atmosphere. This may cause further injuries.

1.2   HOW TO USE THIS BOOK

Failure to identify hazards can lead to undesirable consequences. Risk reduction begins with hazard identification. However, this is just the first step - multiple tools (methods and techniques) must be used to broaden and deepen hazard awareness and control.

While other books cover physical or occupational hazards, this book extends hazard recognition practices and techniques to include identification and mitigation of *process hazards*. One way in which process hazards differ from physical or occupational hazards is that process hazards have the potential to impact more people and the surrounding community. Process hazards are frequently the result of equipment or process systems operating outside their design intent.

Each Chapter provides practical examples that can be used in the field.

Throughout this text, tables, illustrations and photographs have been used to describe various hazards and their effects.

Be a leader, and do not assume others have already identified the hazards around you.

While it may appear useful to have highlighted all of the possible hazards and associated consequences in each and every case, this has not been done. In certain cases, the most common hazards have been explored in detail. In others, more subtle hazards and their potential consequences have been described. However, in all cases the reader is encouraged to identify hazards and consequences beyond those provided in the book.

The interpretation of specific hazards will vary across industry and is generally a function of the types of processes employed, the materials used, and the safeguards in place. For example, the presence of water in a steel mill may be deemed an extreme hazard if it comes into contact with hot or molten metal.

It is difficult to compile a complete listing of all hazards and their range of effects given the wide audience for which this book is intended. Examples of hazards, their effects, and potential consequences are suggested throughout the book to provide examples and to encourage the reader to think beyond one single hazard for each photograph, and one single consequence for each hazard.

To enable your organization to gain the most benefit from this book, a few suggestions for use follow:

- Introduce "hazard recognition" as a standard agenda item at operations safety team meetings.
- Have a supervisor, trainer or worker read the book and then select specific topics for review or training with operations and maintenance staff.
- Select examples from the book to discuss at shift change and other important work group meetings.
- Select a specific hazard to focus on at shift change, tailgate, and similar meetings.
- Flag a page or example in the book and then pass around the control room with a sign-off sheet.

1.3   REFERENCES

1-1   Recommendations for Addressing Recurring Chemical Incidents at the U.S. Department of Energy.

http://www.hss.energy.gov/HealthSafety/WSHP/chem_safety/cstcchemsafetyrecommend.pdf

2

BASIC CONCEPTS

Failure to identify, eliminate or mitigate hazards can have catastrophic results. Identifying and addressing hazards in our workplace helps us go home safely every day.

An effective hazard identification program will help:

- Prevent injuries
- Prevent property damage
- Protect the environment
- Improve business performance, image and reputation

A *hazard* is a condition that has the potential for causing harm to people, property, or the environment. Hazard identification has two key components:

- Identification of conditions in your plant, process, or materials that are hazardous
- Identification of specific undesirable consequences as a result of exposure to the hazard

Hazard identification is an essential step in the elimination or mitigation of workplace hazards.

> **Remember that a hazard must be identified before it can be effectively managed.**

2.1   HAZARD AND RISK

A lot of different words are used when talking about "hazards" and "risk" - and sometimes these words are not used correctly. The words "hazard" and "risk" are often used interchangeably, when in fact they are very different. The European Chemical Industry Council's brochure, "Risk and hazard - How they differ" (Ref. 2-1), does an excellent job of explaining the terms we use to describe hazard and risk. The following definitions and examples are adapted from this publication:

- **Hazard** - *The way in which an object or a situation may cause harm*: A hazard exists where an object or situation has the ability to cause harm. Such hazards include uneven pavement, unguarded machinery, an icy road, a fire, an explosion and a sudden escape of toxic gas. When a hazard triggers a series of escalating hazards, these escalating hazards are referred to as *cascading hazards*.

- **Risk** - *the combination of the likelihood of harm and the severity of that harm.* Risks are all around us in our daily lives. We all conduct risk assessments constantly, whether consciously or sub-consciously. When deciding whether to cross the road, whether to eat healthy food, or what sports to participate in, we make judgments about the hazards involved and assess the risk before taking action. Just as there are risks in our everyday lives, so there are risks in activities that companies perform and in the products they make.
- **Dangerous** - *Something that is inherently or naturally unsafe.* Materials, objects and work tasks can be dangerous to workers.
- **Exposure Potential** - *The likelihood that the hazard will result in harmful consequences:* Safeguards or protection layers will influence the potential for the hazard to result in harmful consequences. The presence of a potential receptor in the area and its distance from the hazard will also influence the extent of the risk. For example, a fire or explosion may cause damage to nearby process structures, but will not harm people if there are no people present at the time. Additionally, if workers are housed in a blast-resistant building, their likelihood of harm from an explosion is minimized.
- **Receptor** - *Any entity that receives or sustains an impact.* Receptors can include people, the environment or physical property.

For risk to be present, there must be a hazard and the potential for exposure to the hazard. Hazard elimination is the only sure way to eliminate risk; however it is possible to reduce risk by minimizing the potential for exposure to the hazard or by containment of the hazard.

The hazard of a chemical refers to its natural properties that may cause an adverse effect for humans or the environment (Ref. 2-1). Risk is the combination of expected frequency of occurrence of harm and the severity of that harm. Hazard identification and evaluation is a key element of managing risks. In order to assess risk, both the hazard and the potential for exposure must be evaluated. Risk assessment provides techniques to evaluate both the severity of the harm and likelihood that it will occur.

2.2   ACCIDENT MODEL

Understanding how incidents occur is critical to recognizing the importance of hazard identification and management. "Domino" (Heinrich, 1931) or "Swiss cheese" (Reason, 1990) accident models are commonly used in the process industries to understand a sequence of failures (chain reaction) that cause an incident or near-miss to occur. This simple type of accident model describes accidents as the result of a sequence of clearly distinguishable events that occur in a specific order. The example in Figure 2-1 depicts several layers of protection that have "holes" - when a set of unique circumstances occur, these "holes" line-up to allow an incident to occur. Protection layers are represented in the figure as "slices of cheese". The holes in the cheese represent failures in the protection layers. These failures can include human errors, mechanical integrity issues that lead to loss of containment or malfunction of a protection layer, management system failures such as failure to recognize and evaluate changes, etc. As the figure illustrates, incidents are typically the result of multiple failures but they begin with the hazard. Similarly, the domino model depicts an accident as a set of dominos that tumble because of a unique initiating event. In this model, the dominos that fall represent the failure of protective layers. These accident models help reinforce the importance of identifying and managing ALL hazards - even hazards that a worker might perceive as "minor".

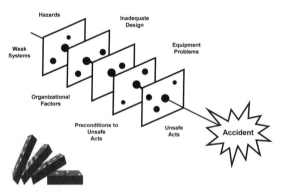

Figure 2-1.   Typical accident models

2.3   PHYSICAL AND PROCESS HAZARDS

Physical hazards are visible and tangible. Exposure to physical hazards typically results in injuries that are a function of energy, area of contact, and duration of contact. Physical hazards are commonly associated with occupational safety hazards and include:

- Slips, trips and falls
- Electrical hazards
- Pinch-points

Physical hazards are often the easiest to recognize - although when they exist at a facility over time, people tend to accept them as normal conditions. For example, when a low hanging pipe crosses an access path in a process unit, worker may accept this as a "normal" condition and no longer recognize it as a "head-knocker" (or as the more severe hazard it could be when evacuating the process area in emergency conditions).

Process hazards are less tangible and can be more difficult to recognize than physical hazards. Process hazards generally have much more significant consequences if they are not mitigated. The potential consequences of process hazards are a function of the type of material involved, its quantity and the process conditions. Process incidents can cause the release of a toxic or flammable material, and may result in subsequent fire, explosion or exposure to toxics. Small process releases can escalate to cause significant injury, environmental impact, or equipment damage.

At some companies, operations and maintenance workers have been trained to identify physical hazards, like the ones pictured on the following pages.

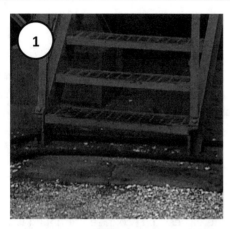

Slips, trips and falls continue to account for many worker injuries. This hazard becomes more significant when combined with a worker attempting to exit an area during emergency conditions.

1) Hoses routed in front of stairs; 2) Opening in catwalk grating; 3) Congested storage blocks exit door.

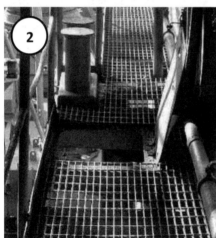

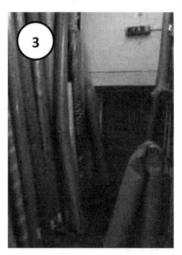

This hazard becomes more significant when it may impact a worker attempting to leave an area during an emergency.

How many examples of these common hazards exist when you do your daily rounds?

What can you do to eliminate them?

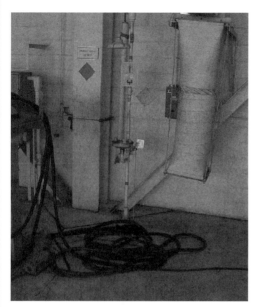

Physical hazards include impeding the ability of personnel to access safety equipment, like safety showers and eye wash stations. Improper maintenance of safety equipment also affects their effectiveness in the event of an emergency.

Impeded access to safety shower

Dirty eyewash station with protective caps removed

Are there areas of your facility where access is impeded?

Do you have housekeeping issues that should be improved?

What can the consequences of these hazards be at your facility?

Electrical hazards are often easy to identify and fix, yet incidents related to exposure to electricity continue to occur. These incidents are typically caused by electrical hazards related to unsafe equipment and/or installation, and unsafe work practices. Electrical components can also be ignition sources when flammable vapors are present. There are various ways of protecting people from the hazards caused by electricity. These include: insulation, guarding, grounding, electrical protective devices, and safe work practices. Higher level protection might include a process to ensure proper electrical code adherence and building/control room siting.

1) Open electrical equipment;
2) Exposed electrical wiring;
3) Exposed electrical wires near fill connection;
4) Open HVAC panel;
5) Congested power lines.

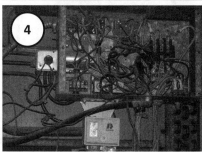

Exposed wiring and unsealed electrical components can provide an ignition source for flammable gases and vapors. In addition, flammable gases and vapors may travel through unsealed conduit to find an ignition source in another area.

What types of electrical hazards do you have at your facility?

What can you do to mitigate or eliminate these hazards?

Improper lifting and/or crane operations expose workers to dropped object hazards. These photos emphasize the importance of protecting workers from the hazards of falling objects and ensuring that those working on the ground are aware of the work going on overhead.

1) Lifting large piping with one sling;
2) Lifting load over people;
3) Lifting pipe over worker below.

What types of work activities are occurring now at your facility that could present hazards like those pictured above?

What can you do to prevent these hazards?   (See color insert)

The physical hazards associated with excavation and trenching operations can be very dangerous. Improper trenching (improper sloping or shoring) leads to cave-ins and worker injury. Inadequate barriers also expose workers to hazards.

1) Open excavation, improper sloping and shoring;
2) Open trench without protective barrier.

Do contractors typically perform trenching activities at your facility?

Who monitors their activities to ensure excavations and trenching are done properly?

What can you do?

Rotating equipment creates hazards that can cause significant injury. Machine guards must be in place to protect worker exposures. Loose clothing should be prohibited around rotating equipment and rings, necklaces, and loose gloves should not be worn.

1) Guard too far away from wheel, glove too loose.

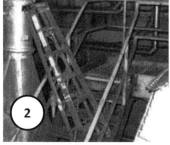

Many ladder accidents are the result of careless or improper ladder usage, making a well-designed and well-trained ladder safety program well worth the effort. This program should include identifying the acceptable ladder classification.

2) Improper use of ladder.

Confined spaces create unique hazards. The consequences of these hazards can be fatal when a worker's ability to exit the confined space is impaired. Small entries also limit the response of rescue workers.

3) Small vessel skirt manway. 12" x 18" on the vertical.

What types of consequences can occur by the three different hazards pictured above? For example, confined spaces can contain oxygen-deficient atmospheres, which can lead to multiple fatalities for those inside the vessel.

Which do you think are the most severe? Unguarded rotating equipment and improper use of a ladder would typically result in an injury to one person, whereas an oxygen-deficient atmosphere in a confined space can result in multiple fatalities.

What can you do to eliminate or mitigate the hazard? Confined space entry should involve continuous monitoring of the atmosphere for sufficient oxygen and hazardous contaminants, along with the use of air-movers, external hole-watches and possible self-contained breathing apparatus.

Improperly supported cable trays can create physical hazards to workers, but they can also escalate to create fire hazards.

1) Improperly supported cable tray.

Some structural hazards are easy to identify and correct, while others take special training to find and evaluate (corrosion of grating, structural steel, etc.)

2) Rusted handrail.

Scaffold hazards cause a significant number of fatalities each year. The vast majority of workers injured in scaffold accidents attributed the accident either to the planking or support giving way, or to the employee slipping or being struck by a falling object.

3) High scaffolding with no access ladder.

How would you identify the hazards depicted above?

Which would be the easiest to identify?

What actions can you take to eliminate or mitigate these hazards? For example, a Planned General Inspection program that makes use of a team of employees to observe and inspect the facility on a periodic basis according to a schedule can be an effective way to identify workplace hazards.

In the simplest form, loss of containment refers to the release of any material from closed process equipment or piping. Loss of containment incidents may also transfer energy in the form of pressure. A worker in close vicinity to a system being depressurized to atmosphere may be seriously injured. Loss of containment may result in:

- Release of flammable material, fire and/or explosion, worker injury and damage to the facility
- Release of toxic material, worker injury, impact to the community

Process hazards can lead to the release of a toxic or flammable material, and subsequent fire, explosion or exposure to toxics. Small events can escalate to cause significant injury, environmental impact, or asset damage. Process hazards can lead to:

- Fires
- Explosions/implosions
- Uncontrolled chemical reactions
- Exposure to:
  - Corrosive materials
  - Toxic materials
  - Ionizing and non-ionizing radiation
  - Pathogens
  - Temperature extremes

These hazardous materials can be solids, liquids or gases. The hazards may be associated with the material size. Fine powders can form explosive atmospheres; liquids can be in the form of droplets or vapors, both of which are generally more hazardous than bulk material (Ref. 2-2 and Ref. 2-3).

Some causes of process hazards may be easy to identify, such as:

- Equipment defects or degradation
  - Corrosion/erosion
  - Impact to piping and equipment
- Inadequate isolation of equipment or piping
- Inadequate energy isolation (lockout/tagout)

Other causes of process hazards are more difficult to detect:

- Design deficiencies
- Deviating from an operating procedure
- Inadequate training
- Operating equipment outside design parameters
- Incorrect MSDS sheets (lack of information can lead to chemical hazards related to chemical instability/reactivity, inadvertent mixing, etc.)
- Equipment not fit-for-service
- Fatigue
- Too many tasks for the current staffing level to perform safely
- Poor communication
  - Between shifts
  - Between equipment and people
- Non-routine operating activities, such as start-up or shutdown
- Test runs
- Feed composition changes
- Swings in ancillary systems, such as utility air, water, fuel gas, inerting gas, etc.

The following photographs provide examples of process hazards.

Vehicular impact to process lines and vessels can have catastrophic consequences and potentially lead to a release of hazardous material, potential fire and explosion and potential personnel exposure and injury. Often, situations as those pictured here, become accepted as a normal condition and the hazards are overlooked.

1) 6" process line vulnerable to vehicle traffic;
2) Solvent spigot not protected;
3) Unsecured gas supply manifold;
4) Building gas manifold not protected from vehicles.

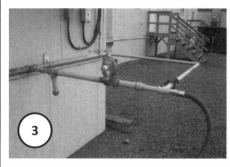

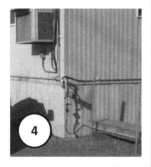

Take another look at the photographs - how many hazards can you identify? Barricade posts are often used to prevent vehicular impact with process equipment. In addition, manifold connections should be capped or plugged when not in use to prevent accidental releases. Often the manifold valves will be equipped with locking-handle designs to further prevent accidental opening of the valve, which can lead to a release of process material.

Are there safety hazards, security concerns or equipment exposure hazards? Or all three?

Corrosion can lead to the loss of containment of a hazardous material, resulting in a fire, explosion or exposure to toxic material.

1) and 2) Corrosion under insulation causes pipe failure; 3) and 5) Corrosion causes valve malfunctions or leaks; 4) Gas line corroded at pipe clamp.

External corrosion can be accelerated by environmental factors (e.g., humidity) or local conditions (e.g., acid vapors venting from an adjacent tank). Report signs of corrosion to your supervisor before the condition results in loss of containment.

What types of corrosion hazards do you have at your facility?

What could the consequences of these hazards be?

What can you do?

What could have been done to prevent the corrosion depicted in the photos?

Is there an effective Preventive Maintenance program that would address problems such as this?   *(See color insert)*

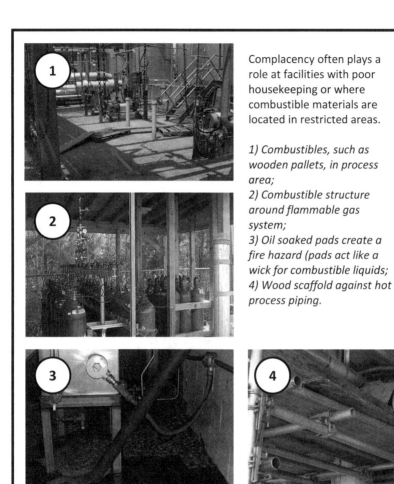

Complacency often plays a role at facilities with poor housekeeping or where combustible materials are located in restricted areas.

1) Combustibles, such as wooden pallets, in process area;
2) Combustible structure around flammable gas system;
3) Oil soaked pads create a fire hazard (pads act like a wick for combustible liquids;
4) Wood scaffold against hot process piping.

Combustible materials, such as wood and oil-soaked absorbent pads, can create a fire hazard, especially when located close to hot surfaces. It is important to look for these conditions when conducting housekeeping inspections and shift rounds and report them to supervision.

What types of housekeeping deficiencies at your facility could lead to process hazards?

Does complacency play a role?

What can you do to eliminate these hazards?

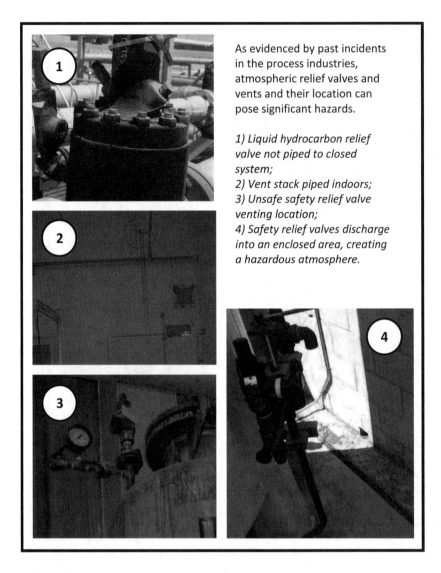

As evidenced by past incidents in the process industries, atmospheric relief valves and vents and their location can pose significant hazards.

1) Liquid hydrocarbon relief valve not piped to closed system;
2) Vent stack piped indoors;
3) Unsafe safety relief valve venting location;
4) Safety relief valves discharge into an enclosed area, creating a hazardous atmosphere.

Process vents, such as pressure relief devices and rupture discs, can create hazards if not vented to safe locations. For example, when flammables are vented indoors or to confined locations, they can result in fires and explosions and personnel exposures.

Do you currently look for these types of hazards at your facility?

How can you incorporate looking for this type of hazard into existing walk-around safety activities?

Repairing operational issues without going through a facility's management of change program can lead to unforeseen process incidents.

1) Pipe flanges used as vibration anchor point;
2) Relief valve (critical safety device) used as vibration anchor.
3) Open piping, no blind flange;
4) Improper isolation. Removal of packing cap and installation of a proper blind flange overlooked.

Even minor changes, such as the installation of support bracing, should be subjected to a Management of Change review to ensure the hazards and risks are addressed.

Have changes occurred at your facility that didn't go through your facility's management of change (MOC) program?

What can you do to ensure that changes are identified and evaluated through MOC?

External factors often play a role in creating hazards in process facilities.

1) Large bird's nest in outlet of safety relief valve;
2) Open vent pipe provides opportunity for plugging;
3) Large bore vent stack without screen resulted in allowing access to vacuum pump outlet.

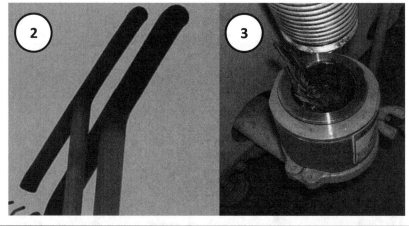

What types of external hazards occur at your facility?

How do you manage these hazards?   *(See color insert)*

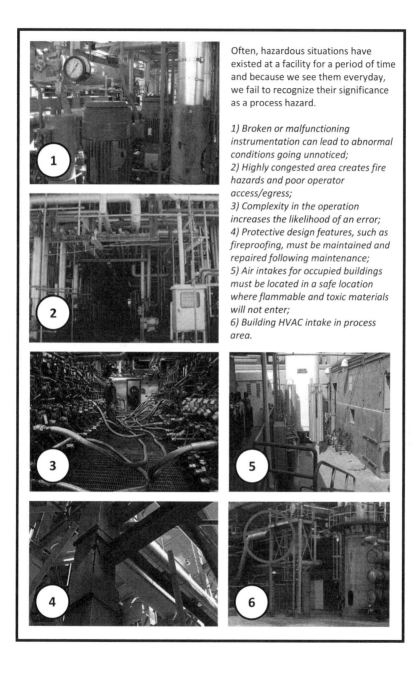

Often, hazardous situations have existed at a facility for a period of time and because we see them everyday, we fail to recognize their significance as a process hazard.

1) Broken or malfunctioning instrumentation can lead to abnormal conditions going unnoticed;
2) Highly congested area creates fire hazards and poor operator access/egress;
3) Complexity in the operation increases the likelihood of an error;
4) Protective design features, such as fireproofing, must be maintained and repaired following maintenance;
5) Air intakes for occupied buildings must be located in a safe location where flammable and toxic materials will not enter;
6) Building HVAC intake in process area.

(See color insert)

2.4   BENEFITS OF HAZARD IDENTIFICATION

The benefits of hazard identification, assessment and control are directly linked to what motivates an employee to champion the program. The more direct the correlation between an employee's hazard management efforts and the benefits to the employee, the more effective the hazard management program. Benefits of hazard identification and management include:

- Enhancing worker safety
- Protecting the public
- Reducing workload (less energy spent recovering from incidents and process upsets)
- Increasing efficiency
- Ensuring job security
- Keeping the company in business
- Improving corporate image

2.5   HAZARDS TYPES BY INDUSTRY

The characteristics of materials/products that make them hazardous are often what make them valuable; consequently, it may be impossible to eliminate all hazards. In these situations, identifying the hazard and applying safeguards is the key to hazard management.

Physical hazards tend to be common at a wide variety of processing facilities. As discussed in Section 2.3, these include:

- Slips, trips and falls
- Electrical hazards
- Pinch-points

Process hazards, although unique to industry types and individual facilities, typically result from loss of containment of a hazardous material. This loss of containment may be caused in several ways, as the photographs in Section 2.3 depicted, but the consequences depend on the type of material, the operating conditions and external factors present at the time of release.

Table 2-1 describes specific process hazards by industry type, focusing on loss of containment of a hazardous material. Other hazards (both physical and process hazards) that are unique to an industry type are described in the column on the far right in Table 2-1.

Table 2-1.   Examples of process hazards in various industries

Industry Type	Loss of Containment Hazards			Reactive Hazards	Other Unique Process Hazards
	Flammable	Explosive	Toxic		
Breweries/ Distilleries	Alcohol presents a flammability hazard.	Grain storage silos present dust explosion hazards.	Anhydrous ammonia is often used for refrigeration. Releases of anhydrous ammonia have the potential to impact both onsite workers and the community.	Inadvertent mixing of incompatible materials.	Asphyxiation hazards when employees enter confined spaces. Asphyxiation from exposure to carbon dioxide.
Chemical Plant (batch)	Leak of flammable liquid, vapors, or dust may lead to fires.	If sufficient quantities of material are released and there is sufficient confinement and congestion, then a vapor cloud explosion hazard may be present. Similarly, if a flammable dust is released, then a dust explosion is possible.	A leak of toxic material could result in either chronic or acute worker exposure leading to injury.	Inadvertent mixing of materials could lead to runaway reactions that result in a rupture of a vessel. Not following the steps of recipe could result in a runaway reaction that could rupture a vessel.	Bulk shipping presents unique hazards.

Table 2-1.   Examples of process hazards in various industries (continued)

Industry Type	Loss of Containment Hazards			Reactive Hazards	Other Unique Process Hazards
	Flammable	Explosive	Toxic		
Chemical Plant (continuous)	Leak of flammable liquid, vapors, or dust may lead to fires.	If sufficient quantities of material are released and there is sufficient confinement and congestion, then a vapor cloud explosion hazard may be present. Similarly, if a flammable dust is released, then a dust explosion is possible.	A leak of toxic material could result in either chronic or acute worker exposure leading to injury.	Inadvertent mixing of materials could lead to runaway reactions that result in a rupture of a vessel.	Vibration can lead to equipment failure and subsequent release. Nitrogen asphyxiation.
Food/ Beverages Processing	Fork-lifts may use propane or diesel as fuel. This presents a hazard to workers when refueling fork-lifts.	A release of anhydrous ammonia in an enclosed area has the potential for explosion. Processing plants such as those for grains (flour) and sugar have a risk of dust explosion.	Anhydrous ammonia is often used for refrigeration. Releases of anhydrous ammonia have the potential to impact both onsite workers and the community.	Materials of construction and repair must be compatible with anhydrous ammonia.	Contamination of product can lead to wide-scale public exposure and health effects, recall of product, and loss of reputation. Carbon dioxide asphyxiation.

Table 2-1.   Examples of process hazards in various industries (continued)

Industry Type	Loss of Containment Hazards			Reactive Hazards	Other Unique Process Hazards
	Flammable	Explosive	Toxic		
Gas Plant	Leak of flammable liquid or vapors may lead to fires.	If sufficient quantities of material are released and there is sufficient confinement and congestion, then a vapor cloud explosion hazard may be present.	The toxic hazards at gas plants are typically limited to hydrogen sulfide and chlorine.	Materials of construction and repair must be compatible with hydrogen sulfide.	Carbon dioxide may be present in some gas plants which can increase corrosion resulting in leaks. Carbon dioxide asphyxiation.
Mining	Flammable fuels used in equipment.	Seepage of methane gas into underground mine presents explosion hazards. Use of explosives in mining process presents unique hazards. Dust explosion of combustible coal dust.	Build-up of carbon dioxide presents asphyxiation hazards. Cyanide. Sulfur dioxide and carbon monoxide may be present in mine shafts (Sulfur dioxide is released during blasting).	Solvents and acids can react with other materials in the process.	Crushing. Mine/highwall collapse. Flooding. Rock burst.

Table 2-1.   Examples of process hazards in various industries (continued)

Industry Type	Loss of Containment Hazards			Reactive Hazards	Other Unique Process Hazards
	Flammable	Explosive	Toxic		
Offshore Facility	Leak of flammable liquid or vapors may lead to fires.	If sufficient quantities of material are released and there is sufficient confinement and congestion, then a vapor cloud explosion hazard may be present.	The toxic hazards at an offshore facility are typically limited to hydrogen sulfide (which may be entrained and subsequently removed from crude oil).	Materials of construction and repair must be compatible with salt water.	Most offshore facilities have some processes which are high pressure (>5,000 psi (345 bar)) that pose unique hazards. Dead leg sections of pipe that are out-of-service or rarely/ infrequently used can be a source of process leakage. Entrained water from the process stream can settle and freeze in the dead leg causing pipe damage. Caution should be taken that water does not collect in dead legs and these sections of piping should be eliminated. Marine operations present loading/ unloading hazards. Emergency response actions are limited because there is no place to go.

Table 2-1.   Examples of process hazards in various industries (continued)

Industry	Loss of Containment Hazards			Reactive	Other Unique
Type	Flammable	Explosive	Toxic	Hazards	Process Hazards
Oil Refinery	Leak of flammable liquid or vapors may lead to fires. Light hydrocarbon and hydrogen.	If sufficient quantities of material are released and there is sufficient confinement and congestion, then a vapor cloud explosion hazard may be present. Light hydrocarbon and hydrogen.	The toxic hazards at a refinery typically include hydrogen sulfide (which may be entrained and subsequently removed from crude oil and natural gas), sulfuric acid or hydrofluoric acid (which are used in Alkylation Units), and chlorine (which may be used to treat cooling water).	Common reactions between bases and acids may exist.	Large tank inventories susceptible to lightening strike and subsequent tank fires. Cogeneration Units present high voltage and high pressure steam hazards. Legionella bacteria in cooling towers presents heath hazard to workers.
Pharm-aceutical	Leak of flammable liquid, vapors, or dust may lead to fires.	Accumulation of dust presents explosion hazards.	A leak of toxic material could result in either chronic or acute worker exposure leading to injury.	Inadvertent mixing of materials could lead to runaway reactions that result in a rupture of a vessel. Not following the steps of recipe could result in a runaway reaction that could rupture a vessel.	Off-spec or contaminated products may cause harm to the general public.

Table 2-1.   Examples of process hazards in various industries (continued)

| Industry | Loss of Containment Hazards | | | Reactive | Other Unique |
Type	Flammable	Explosive	Toxic	Hazards	Process Hazards
Pipelines	Leak of flammable liquid or vapors may lead to fires.	Overpressure from blocked-in pipelines or compressor stations may lead to rupture and explosion.	A leak of toxic material could result in either chronic or acute worker exposure leading to injury.	Corrosion inhibitor chemicals not compatible with materials of construction. Materials of construction and repair not compatible with hydrogen sulfide.	In-line scraping operations can expose workers to high pressure releases. Dead leg sections of pipe that are out-of-service or rarely/infrequently used can be a source of process leakage. Entrained water from the process stream can settle and freeze in the dead leg causing pipe damage. Caution should be taken that water does not collect in dead legs and these sections of piping should be eliminated. An undetected leak is a danger to the environment.
Power Generation (including nuclear)	Leak of flammable liquid or vapors may lead to fires.	Coal dust presents explosion hazard.	Anhydrous ammonia may be used for nitrogen oxides control at Co-Gen facilities. Releases of anhydrous ammonia have the potential to impact both onsite workers and the community.	Runaway reaction potential exists in nuclear facilities. Nuclear radiation.	Electrical hazards are magnified at these high voltage facilities. Fly ash and other solid wastes blowing into community. Nuclear waste presents radioactive hazards. Carbon monoxide toxicity issue.

Table 2-1.   Examples of process hazards in various industries (continued)

Industry Type	Loss of Containment Hazards			Reactive Hazards	Other Unique Process Hazards
	Flammable	Explosive	Toxic		
Pulp and Paper	Combustible dust may lead to fires. Turpentine. Non-condensable gases.	If a combustible dust is released, then a dust explosion is possible. Paper dust. Sawdust.	Chlorine, chlorine dioxide and sulfur dioxide are both used in the processes and have the potential to impact both onsite workers and the community.	Chlorine is a strong oxidizer and will react with most metals and organic materials. Also, chlorine reacts with water, creating the potential for worker exposure to hydrochloric acid.	Physical impact of slurry viscosity. Low pH of water causes topical exposure issues. Conveyers, moving parts and heavy loads. Wood chippers provide exposure hazards to personnel.
Steel Manufact-uring (and other metals)	Loss of containment of molten metals above auto-ignition temperatures of contacted materials results in subsequent fires.	Water contacting high temperature metals can result in a steam explosion.	Gases from smelting operation contain toxic materials.	Solvents and acids can react with other materials in the process.	High-heat manufacturing process presents worker exposure issues.
Upstream Oil and Gas Facility	Leak of flammable liquid or vapors may lead to fires.	If sufficient quantities of material are released and there is sufficient confinement and congestion, then a vapor cloud explosion hazard may be present.	The toxic hazards at an upstream facility are typically limited to hydrogen sulfide (which may be entrained and subsequently removed from crude oil).	Materials of construction and repair must be compatible with hydrogen sulfide.	Facilities may have some processes which are high pressure (> 5,000 psi (345 bar)) that pose unique hazards. Diesel trucks and generators in proximity to light end hydrocarbon vapor. Well-blowouts present unique hazards.

Table 2-1.   Examples of process hazards in various industries (continued)

Industry Type	Loss of Containment Hazards			Reactive Hazards	Other Unique Process Hazards
	Flammable	Explosive	Toxic		
Water and Wastewater Treatment	Diesel driven water pumps present a potential fire hazard. Storage of dry sludge may contain pyrophoric iron sulfides.	Methane gas accumulation and its subsequent ignition presents explosion hazards.	Chlorine, sulfur dioxide, and anhydrous ammonia can be used at facilities that treat water and wastewater. Releases of these materials have the potential to impact both onsite workers and the community.	Chlorine and sulfur dioxide are strong oxidizers and will react with most metals and organic materials. Sulfuric acid will react with concrete and produce a reactive by-product.	Exposure to contaminated water may cause adverse health effects. There has been a history of worker injury and fatality due to unsafe confined space entry activities in open pit treatment areas.

This table presents examples of hazards by industry type, however there are hazards that this table does not identify.

How many hazards at your facility are not identified here?

Here's a hazard unique to the mining industry (or other industries that use heavy haul trucks):

WARNING – HEAVY HAUL TRUCKS

Heavy haul trucks are used in the mining industry to transport ore. Over the years these have increased in size with some having a load capacity approaching 400 tons. The 12 ft diameter tires of these vehicles are typically inflated to 90 psig and are mounted on multi-rim wheels. When the tires are serviced the rims must be removed in sections; this requires the loosening and removal of as many as 60 bolts.

Given the immense size of these tires, even the slightest residual pressure (remaining in a tire) can create a large outward force pushing the tire off the rim. This can kill or seriously injure nearby workers. Mining operations worldwide have frequently experienced such incidents. When maintaining pressurized equipment, always bleed the system pressure to zero and stand out of the direct line of fire. Use a protective safety cage when re-assembling the wheel. This principle should be applied when dealing with all pressurized equipment.

There is also the concern of haul truck tire explosions caused by tires overheating, due to low pressure. The tire rubber may decompose and create a flammable mixture in the tire. Workers trying to change the tire may strike a spark with metal tools on the rim and the tire may explode. This is why many mines use nitrogen to pressurize heavy equipment tires.

2.6 REFERENCES

2-1. Cefic - The European Chemical Industry Council, *Risk and hazard - How they differ*, August, 2003.

2-2. Imperial Sugar Company Explosion and Fire, Port Wentworth, GA. Current CSB Investigation. www.csb.gov.

2-3. Silver Eagle Refinery Flash Fire, Woods Cross, UT. Current CSB Investigation. www.csb.gov.

3

IDENTIFY HAZARDS

Identifying hazards is the first step in a hazard management process (Figure 3-1).

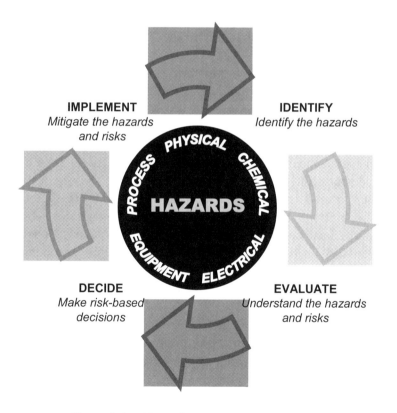

Figure 3-1.   Hazard management process

A facility's culture and management practices influence a hazard management program's effectiveness:

- **Complacency:** A facility's culture or personal beliefs can influence whether or not an individual believes there is the potential for "something to go wrong".   Often, the mistaken belief that nothing can go wrong is influenced by "nothing has ever happened before" or acceptance of the hazard as a normal working condition.   To prevent complacency, it is important to maintain a sense of vulnerability.
- **Management Commitment and Empowering Workers:** A facility's commitment to providing resources and training for hazard identification – and then doing something to eliminate or mitigate the hazard once it is identified – dramatically strengthens a hazard management program.

Given an environment of management support and encouragement in the hazard management process, our individual ability to recognize hazards is influenced by our senses. We will begin by understanding our senses - vision, hearing, touch, smell and taste (Figure 3-2) - which allow us to be aware of the conditions around us. Develop your five senses as a skill set. These are your front line defenses and when used correctly greatly enhance your ability to identify hazards.   However, the use of the senses must not be used as a substitute for established and more rigorous hazard detection methods (i.e., gas detection, vibration monitoring, etc.).   Over time, workers can become desensitized to normal conditions. A new worker or visitor might be in a better position to identify an unusual condition based on experience elsewhere. It is important to realize that by the time your senses recognize the danger, safe exposure levels may have already been exceeded.

> **Sensory detection is not a substitute for**
> **established hazard detection and control measures.**

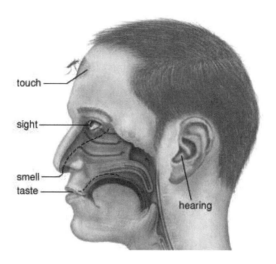

Figure 3-2.   Five basic senses

It is important for us to appreciate the capabilities and limitations of our senses so we can better respond to the hazards we face:

- Vision is by far the most important of our senses since almost 90 percent of the hazards we encounter are visual. We need to know, for example, that we can't see as well in low levels of illumination such as night time conditions or under high levels of illumination such as glare.
- Hearing is the next most important sense. But hearing is adversely affected by high levels of background noise.
- Touch, smell and taste can warn us of a spill, a sharp object, or the release of a toxic chemical. But these senses are often imprecise and the hazard of concern may be below our level of detection.

We should also understand that the senses of some people are more acute than those of others. So while one person may be able to hear the hiss of gas escaping from a line, another person might not.

What senses could you use to identify the hazard depicted below? For example, a small leak from a drum may be hidden from sight, but can create a detectable odor. Similarly, problems inside process equipment may not be visible, but may create unusual sounds that are noticed by operators.

Using our senses can help detect hazards that can have catastrophic consequences.

Once we have detected a potential hazard, we need to appreciate its likely effects on us and those around us. We do this mostly through experience. Fire, for example, may not be recognized as a hazard the first time we encounter it (Figure 3-3).

But once we have felt its effect on our body, we learn to respect the hazard. Our experience is centered in our memory. Later in this chapter we will discuss how higher order processes, like short- and long-term memory, help us to appreciate the hazards we encounter.

Figure 3-3.   Experiencing a hazard for the first time

INCIDENT – FLASH FIRE

A pinhole leak was discovered on an elevated pipe at an oil refinery. The leaking carbon steel line was 6" in diameter, operated at 12 psig and contained naphtha (a highly flammable liquid). In an effort to correct the problem, the pressure was slightly reduced and an epoxy plug was inserted into the hole.

The leak continued and a subsequent inspection revealed that the entire pipe circuit was severely corroded and approaching failure. This would require that the line be replaced. An aggressive work plan was developed to replace the line with the process unit in service. Several workers advised that the unit should be shutdown to carry out this work. Nonetheless, the job proceeded. An attempt was made to block in the line at both ends using the existing gate valves. The closed valves appeared to provide isolation but the threaded stem on at least one valve protruded well into the open position; this should have served as a warning. After two weeks of planning and several failed attempts to remove the corroded piping, the line was finally cut in two places. A sudden surge of naphtha splashed onto several workers and immediately ignited. Four workers were killed and one was critically injured. The unit sustained considerable fire damage.

How could using your senses have prevented this incident? Looking at the valve stem should have shown that the valve was not closed. What other indications might you have noticed?

3.1   CONCEPT OF RECOGNITION

Recognition is a process that compares a representation already in short-term or long-term memory with a condition or circumstance in the real world.

It is generally agreed that humans process information in a sequential fashion according to the information processing model shown in Figure 3-4. The model shows a three stage activity: perception, cognition and response selection (Ref. 3.6). Each activity can be further divided into basic functions that operate as filters, processors and storage in both closed- and open-loop fashions as shown.

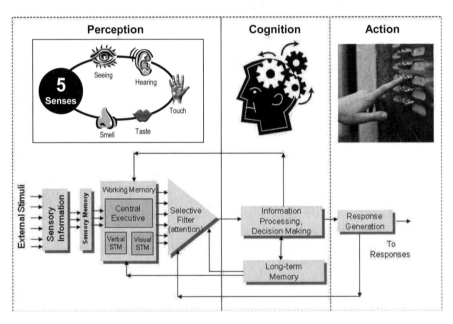

Figure 3-4.   An information processing model

Detection requires the hazard to be capable of being sensed (seen, heard, smelled, tasted or touched) by the worker. Section 3.2 examines the properties of basic human senses that allow an object to be detected. Even though an object may be detectable, the worker may not pay attention to it. Section 3.3 discusses attention, how attention may be disrupted and how it may be improved.

Once an object has been attended to, it is recognized by comparing it with memory, decisions are made and stored in memory, and responses are selected and generated. All of this can happen in as little as 1/5 of a second. The effects of memory on recognition and recall are discussed in Section 3.3.

The concept of "Hazard Recognition" implies that the hazard to be recognized is a "condition or situation that can cause harm to people and damage to property and environment when the hazards are not controlled" (Ref. 3-1). The hazard may be physical (e.g., exposed, sharp, aluminum covering for pipe insulation) or a set of circumstances (e.g., the process of connecting a hose to a high pressure utility air line). In most cases, the hazard is learned from the experience of others, from procedures, referenced in training sessions, or identified in plant surveys, rather than personally experienced.

3.2   BASIC HUMAN SENSES

The relationship between human senses, mental processing and response can be explained by examining the task of the fork-lift driver in 0. The driver receives information from his working environment. In this case, it is the location of the load that he has to pick up and transport. He receives the information visually, compares it with stored experiences, and responds by using the fork-lift controls. His ability to pick up the load without error may be impaired by fatigue, medications, level of experience, workload or stress. The driver's control response may include stepping on a pedal or moving a control lever. This response, in turn, modifies characteristics of the job (e.g., the load is lifted). The load's new position is once again seen by the worker who processes the changed information and initiates a new response (e.g., backing the fork-lift with the load.)

Figure 3-5.   Fork-lift driver coordinates his senses, mental processing and responses to safely transport a load

What are the hazards associated with this job task? For example, an imbalanced load may fall off of the forks or the forklift may be accidentally driven into process piping, resulting in loss of containment.

What could happen if the hazards are ignored? How can you use your senses to detect the hazard? What can you do about it?

The next section reviews the basic characteristics of human senses that permit the worker to receive information, and, ultimately, recognize hazards. This chapter discusses each sense, in order of importance.

Hazards may be seen far away.
Hazards are much closer if you can hear them.
If you can smell, touch or feel the hazard, response must be immediate.

3.2.1 Vision

Most of the information we receive is through our eyes. Visual information can be in the forms shown in Figure 3-6. Basically, our eyes act like a video camera. Light enters through a lens and is focused on a photosensitive area where it is converted to an electrical signal which passes to our brain.

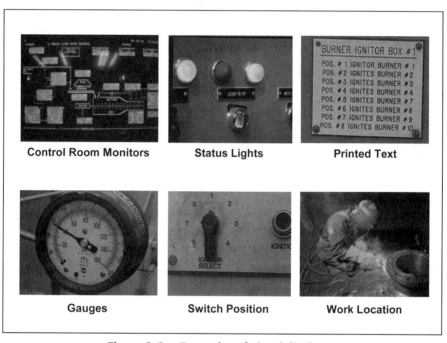

Figure 3-6. Examples of visual displays

Our eyes are able to function over a wide range of illumination levels. However, several external factors also determine visual capabilities and effectiveness, creating important implications in detecting potential hazards. These factors are discussed below.

3.2.1.1 Visual Clarity

As we age, our eyes lose their elasticity and our ability to focus our eyes is affected. When this occurs, two conditions are common:

1. We cannot focus on far objects
2. We cannot focus on near objects

When we are unable to focus properly, our vision is blurred (Figure 3-7). Blurred vision can be corrected with glasses or contact lenses.

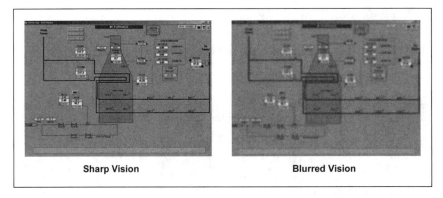

| Sharp Vision | Blurred Vision |

Figure 3-7.   Sharp and blurred images of a CRT display

3.2.1.2 Visual Field

If we are not looking at an object, we cannot see it. So, we continually scan the visual scene by moving our eyes and our head to see objects around us. This is because our angle of sharp vision is limited to those objects directly in front of our eyes. So, it's important that those things we need to see are not located off to the side or behind us. If they are, we might not be able to see them.   Consider, for example, a large control panel (Figure 3-8). Important displays (such as alarms, Figure 3-9) should be located in the center of the console for easy detection.

Figure 3-8.   Wide process control console

Figure 3-9.   Wide control panel with alarms

3.2.1.3 Ability to Adjust to Brightness

Each time we transition between a dark visual environment and a bright one, our eyes have to adjust to the change in illumination. The adjustment takes time, during which we may not be able to focus on objects. We transition from dark to light much faster than from light to dark. It can take as much as 30 minutes to adapt to a very dark visual environment from a very bright one. So, workers need to:

- Allow sufficient time for the eyes to adapt when performing tasks at low levels of lighting
- Avoid excessive changes in illumination levels since the constant need for adaptation results in visual fatigue
- Avoid excessive differences in brightness within the visual field

3.2.1.4 Color Vision

In the workplace, color is used to code equipment and facilities, helping the worker distinguish items. Emergency shutdown switches, for example, are often coded 'red', and green-colored valves on a control screen mean the valves are open. Yellow is used on CAUTION signs to help workers understand that a hazard may exist. As noted above, the ability of the eyes to distinguish color reduces as the level of illumination decreases. In nighttime conditions, for example, colors disappear as each item becomes a shade of gray. It is important to consider the lighting conditions under which the items of interest (e.g., color coded displays or switches) will be viewed.

People with normal color vision are capable of distinguishing hundreds of different colors. Complete lack of color vision (color blindness) is extremely rare. Color deficiency, such as the inability to discriminate between red and green is quite common. Usually, these colors are seen as poorly saturated yellows and browns. Overall, about 8% of the male population and less than 1% of the female population have this form of color deficiency. And, about 1% of the population is unable to accurately distinguish between yellow and blue. It's important, therefore, that color is used as a redundant (additional) method of coding, and not as the primary method.

From this distance, would a person with a color deficiency be able to determine the alarm conditions on this panel? (See color insert)

What other hazards can be identified here? (Hint: nitrogen purge on panel). Nitrogen is an asphyxiant gas, as well as being odorless and colorless. Nitrogen leaks can be virtually impossible to detect with the senses, and can very rapidly create oxygen-deficient atmospheres. In addition, the use of nitrogen as a purge gas for this electrical panel suggests the presence of flammable materials in the area.

3.2.1.5 Resolving Detail

Even though an item may be visible, we may not be able to see the detail in it. For example, we may be able to see a gauge, but not be able to read it. Visual acuity is the term used to describe seeing detail in an object (Ref. 3-2). Some examples are shown in Figure 3-10.

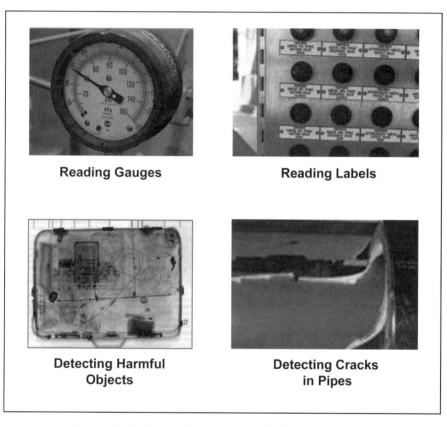

Figure 3-10. Examples of the need to see the detail

Our ability to see detail can be reduced by inability to focus, color deficiency, and transitioning between light and dark environments. But, acuity is most affected by situational factors which include the size of the object, the contrast between the object and its background, and the level of ambient illumination.

3.2.1.6 <u>Size of Object (Visual Angle)</u>

Figure 3-11 illustrates that, for a given viewing distance, small letters are harder to read than big letters. As the size of letters increases or as we move closer to them, our ability to see them improves.

Small letters are hard to read

Big letters are easy to read

Figure 3-11.   Illustrating the effects of character size on reading ability

If a sign is designed for a plant or an annunciator for a control panel, the maximum viewing distance must be considered, and the character height must be adjusted accordingly. The relationship between visual angle, viewing distance, and object size is illustrated in Figure 3-12.

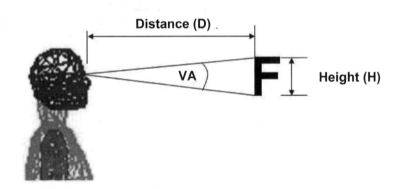

Figure 3-12.   Ability to read a character depends on object size and distance from the eye

Under favorable viewing conditions - good lighting, good contrast, non-critical displays

$$H = D/200 \qquad (Ref. 3\text{-}2)$$

Where: H = character height in meters

D = viewing distance in meters

Under unfavorable viewing conditions - low light levels, reduced contrast, critical displays

$$H = D/120 \qquad (Ref. 3\text{-}3)$$

For example, the size of letters to be used for a sign that summarizes evacuation instructions and must be clearly visible from 10 meters (30 feet) away under dim lighting conditions is:

$$H = D/120$$
$$= 10/120$$
$$= 0.08 \text{ m or } 0.08 \text{ cm high } (3.15 \text{ in})$$

For another example of the importance of acuity, consider Figure 3-13.

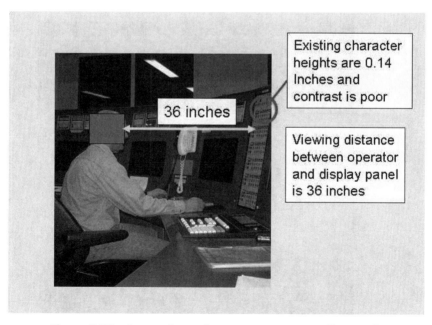

Figure 3-13. Annunciator characters are too small to read

This control room operator must be able to read the words on the annunciator tiles located on his control panel. If the operator's viewing distance is 36 inches and the contrast between the tile letters and background is poor, the minimum height of the letters, as determined by Equation 2 above, is:

$$H = 36/120$$
$$= 0.3 \text{ inches}$$

3.2.1.7 Contrast

Contrast is a measure of how well an object stands out from its background. The more contrast between an object and its background, the easier it is to see. White letters against a black background, for example, are much easier to see than dark gray letters against the same background (Figure 3-14).

Dark letters on a dark background are hard to read

Dark letters on a light background are easy to read

LIGHT letters on a DARK background are easy to read

Figure 3-14. Illustrating the effects of contrast on reading ability

Additionally, light letters on a dark background are more easily obscured by industrial dirt and grime than dark letters on a light background.

As a practical example of the effects of contrast on reading ability, Figure 3-15 shows three labels:

- Black letters on a brown background
- Black letters on a silver background
- Black letters on a red/orange background

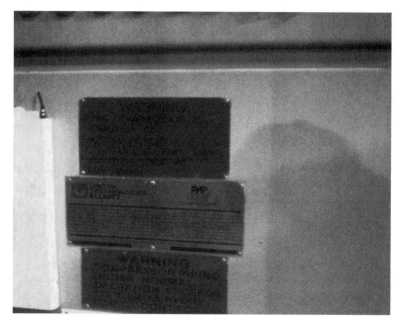

Figure 3-15. Example of poor contrast in safety signage

What are the hazards identified in this picture? The poor contrasting on the signs creates a hazard. Workers may not recognize or be able to read the warning labels.

What can happen if the hazards are ignored? What can you do about it?

Color contrast works in much the same way. The darker the color, the more difficult it is to see against a dark background. Table 3-1 compares the acceptability of combining colors with different contrasts with each other (Ref. 3-2).

Table 3-1.   Acceptability of combining various colors as objects and backgrounds

Color	Black	Dark Blue	Dark Red	Pink	Light Orange	Lime Green	Cyan	Yellow	White
Black (3)									
Dark Blue (10)									
Dark Red (20)									
Pink (30)									
Light Orange (40)									
Lime Green (50)									
Cyan (60)									
Yellow (70)									
White (98)									

Unacceptable	Marginally Acceptable	Acceptable

3.2.1.8 Illumination

As the light level increases, so does our ability to see detail in an object. The caution is that very high light levels can also cause glare. The ideal light level is determined by the activity that is being performed. Generally speaking, the more intense the visual task, the higher the light level that is required. So, for example, it doesn't require as much light to walk down a hallway as it does to change the brushes in a motor. For process plants, recommended light levels are given by Table 3-2 (Ref. 3-3).

As a practical guide, if your work area seems like it is too dark, the light levels should be measured. If you think it's too dark, it probably is. If you can't see clearly, there is a potential hazard. When assessing visibility and illumination, it is important to consider a dark night with inclement weather conditions.

Table 3-2.   Recommended illumination levels by process area (from Attwood et al. 2004)

Process Area	Illumination (lux)
Control Room	
Desk area and console	325
General	215
Manual Sampling Points	215
Compressor Houses	160
Loading Racks	
Loading point	110
General area	32
Platforms	
Main operating area	55
Ordinary area	22
Stairways and Ladders (frequently used)	55
Exchanger Areas	32
Streets and Parking Lots	4

Most hazards are detected with vision. Table 3-3 provides a few examples of hazard indicators that are detected visually, the potential cause and hazard effects, and potential solutions.

Table 3-3.   Examples of visual hazard indicators

Visual Hazard Indicators	Possible Cause	Hazard Effects	Factor that can Impair Detection	Potential Solution
Smoke	Loss of containment	- Fire - Explosion - Equipment damage - Inability to escape - Workers injury	- Low levels of illumination	- Light area that may show smoke
Movement	Rotating equipment	- Toxic exposure - Fire/explosion - Mechanical injury	- Low illumination - Inability to focus	- Increase illumination - Improved corrected vision for workers
Color coded equipment	High voltage Emergency signal cables	- Electrocution	- Poor acuity - Low levels of illumination - Poor color contrast - Poor color rendered lighting	- Increase size, contrast of signal to background - Increase illumination - Modify color of lighting
Color coded lights	Toxic atmosphere area, outlining physical hazard area	- Poisoning - Asphyxiation - Mechanical injury	- Low levels of brightness - Incorrect color coding	- Increase brightness of signal lighting - Improve color coding
Ground reflection	Spilled liquid	- Slips	- Low levels of illumination - Poor contrast between liquid and surroundings	- Increase illumination - Improve background contrast
Corrosion	Mechanical failure	- Property damage - Physical injury	- Poor contrast - Low illumination	- Improve background contrast - Improve illumination

Table 3-3.   Examples of visual hazard indicators (continued)

Visual Hazard Indicators	Possible Cause	Hazard Effects	Factor that can Impair Detection	Potential Solution
Exposed electrical cables	High voltage	- Electrocution	- Poor contrast - Low levels of illumination	- Protect cables from accidental contact - Increase color contrast - Increase illumination
Lines or cables across walking path	Trip	- Physical injury	- Poor acuity - Poor contrast between line/ cable and background - Low illumination	- Increase contrast - Increase illumination
Wear marks	Mechanical failure	- Physical injury	- Poor contrast	- Increase contrast - Increase illumination
Safety signage	Multiple hazards	- Multiple results	- Signs too small - Signs wrong color	- Increase character sizes - Correct color coding
Obstructions	Physical contact	- Head/body trauma	- Poor contrast	- Increase contrast - Increase illumination
Steam lines	Physical contact	- Burns	- Poor contrast (color coding)	- Improve color contrast
Exposed insulation	Physical contact	- Cuts	- Poor contrast - Poor lighting	- Repair insulation - Improve contrast - Increase illumination
Snow, ice, rain	Trip, vehicle impact	- Physical injury	- Poor lighting	- Increase area illumination - Improve clean-up and drainage

Table 3-3.   Examples of visual hazard indicators (continued)

Visual Hazard Indicators	Possible Cause	Hazard Effects	Factor that can Impair Detection	Potential Solution
Field control panel	Unreadable displays	- Process upset	- Glare - Acuity - Color coding	- Reduce/ eliminate glare - Increase panel display intensity - Reorient panel - Block glare source

3.2.1.9 Hazard Signage

Hazard signs serve two major purposes in a facility and should take into account the vision limitations and abilities discussed above. Hazard signs identify equipment and facilities that require special attention. Additionally, they serve a training function for new workers by advising them of the equipment and facilities that demand safe behavior and whose use is likely covered by a critical operating procedure. The design of HAZARD identification signs follows the American National Standard Institute (ANSI) Z535.4-2002 (Ref. 3-4) and International Standards Organization (ISO) 3864-1:2002 (Ref. 3-5) standards.

Standards generally specify three major levels of HAZARD Identification Signage:

- DANGER: Indicates an immediately hazardous situation which, if not avoided, will result in death or serious injury
- WARNING: Indicates a potentially hazardous situation which, if not avoided, could result in death or serious injury
- CAUTION: Indicates a potentially hazardous situation which, if not avoided, may result in minor or moderate injury. It may also be used to alert against unsafe practices.

Figure 3-16, Figure 3-17 and Figure 3-18 illustrate examples of each of the three major levels of hazard signage and their design requirements.

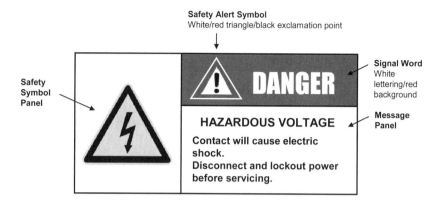

Figure 3-16.   Example of DANGER hazard sign *(See color insert)*

Figure 3-17.   Example of WARNING hazard sign *(See color insert)*

Figure 3-18.   Example of CAUTION hazard sign *(See color insert)*

The sizes of the signs, their graphics and their characters will vary, depending on how far away they must be read. Typically, character sizes for the signal words will be larger than those for the other text on the sign as follows:

SIGNAL WORDS (e.g., DANGER)

$$H = D/120$$

Message Panel Characters

$$H = D/150$$

Where:

H = Character height

D = Reading distance

Hazard signs are mounted at, or immediately adjacent to, sources of hazards and at the point where the hazard could be initiated. The hazards for which signs are typically provided include:

- Radiation hazards
- Mechanical and electrical hazards
- Noise hazards
- Chemical hazards
- Physical hazards

3.2.2   Hearing

Compared to vision, which provides about 90% of the information people process, hearing provides about 5%. Even so, hearing is critical for recognizing hazards. Field workers talk to others in the plant, directly and by radio. Emergency alarms warn workers of hazardous conditions. Workers can often sense how well equipment is operating by its sound. Hazard identification is affected by two auditory issues:

- Expectation
- Hearing sensitivity

Workers learn through experience that certain sounds in a process plant are normal and others are abnormal. Normal sounds include the hum of an electrical transformer or the hiss of product being transferred from one line to a vessel. Abnormal sounds often indicate an impending hazard of a release or electrical explosion. Table 3-4 lists some examples of audible hazard indicators.

Table 3-4.   Examples of audible hazard indicators

Audible Hazard Indicators	Possible Cause	Indicates	Hazard Effects
Hissing	Pipe or equipment leak	Loss of containment	Toxic exposure Explosive cloud Steam exposure
Siren	Plant upset	Loss of containment	Explosion Fire Toxic release
Mechanical banging	Rotating equipment imbalance	Equipment damage Loss of containment	Hydrocarbon release
Line banging	Unstable control valve Line cavitations Line blockage	Loss of containment	Line rupture and hydrocarbon/steam/toxic release
Mechanical screaming	Pressure relief valve release	Loss of containment	Hydrocarbon/steam/ toxic release
Buzzing/hum	Electrical arcing	Electrical exposure to workers	Electrical damage Worker injury
Machine cycling - change in sound pattern	Mechanical damage	Loss of containment	Worker injury Shrapnel

Two characteristics of a sound determine how well we can hear it. The first is loudness. If the sound is too quiet, we may not be able to hear it. If it's too loud, the sound could damage our hearing. Although most corrosion is detected visually, if corrosion is undetected a leak from a corroded pipe may be detected by hearing the release (Figure 3-19).

Figure 3-19. Corroded pipe

What could the consequences of this hazard be? Severe corrosion can lead to unnoticed pinhole leaks or possibly catastrophic failures at high pressures.

Do you think this type of hazard goes undetected or is ignored?
Often, corrosion is hidden from sight, such as underneath insulation, on the underside of piping, or at the base of storage tanks.

What can you do about it? Look for signs of corrosion and report these observations to supervision.

Table 3-5 lists the sound pressure levels (loudness) of typical industrial sounds (Ref. 3-1). Under ideal living and working conditions, the range of sound pressure levels would vary between 20 and 60 decibels (dB). Sounds below 20dB are difficult to hear (depending on the frequency) and those above 60dB are annoying. The human ear can be damaged by continuous exposure to sound levels of 85dB or higher (Ref. 3-6). Thus, noise above 85dB is considered an industrial hazard. Hearing damage increases with increasingly high sound pressure levels and as the length of exposure to high sound pressure levels increases.

Table 3-5. Peak sound pressure levels of common sounds in decibels

Source	Sound Pressure Levels in Decibels
Motor test bench	130
Pneumatic bore-hammer	120
Release of a relief device	115-120
Rocking sieve; chain saw; compressed air riveter; electric cutter; compressed air hammer	105-115
Fin-fans (air cooled exchanger)	100
Furnace air (naturally induced)	90
Tool making machine (running light)	80
Average busy office	45-60
Quiet conversation	30

Decibels are units for expressing the intensity of sound on a scale from zero (for the average least perceptible sound) to about 130 for the average pain level.

The second characteristic of a sound that determines whether we can hear it is the frequency, or pitch, of the sound. Humans can hear lower frequency sounds (e.g., human voice) better than higher frequency ones (Ref. 3-7). Figure 3-20 illustrates that as we age, we continually lose our ability to hear high frequency sounds.

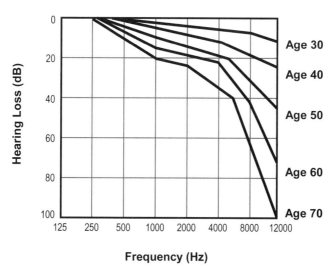

Figure 3-20. Distribution of hearing loss by age and frequency

Noise is defined as unwanted sound. Excessive noise can:

- Damage hearing
- Interfere with speech
- Mask sounds relevant to job performance, thus increasing the difficulty of the job
- Affect performance in tasks requiring concentration
- Mask warning signals
- Cause considerable annoyance and stress

Moreover, the "startle response" from a sudden loud noise can cause muscle contractions, head-jerk movements, blinking and can affect perceptual-motor performance.

Noise is best controlled by treating it at its source. Methods of reduction involve (Ref. 3-8):

- Reducing the sound pressure level or altering the frequency spectrum of the generated noise (proper design and maintenance of machinery, use of rubber dampers, isolating vibrating parts, etc.)
- Using barriers to block or attenuate noise transmission through the air or structures (high frequency noise is better contained by barriers than low frequency noise). Barriers can be supplemented with the use of personal protective devices such as ear muffs and plugs.
- Absorbing direct or reflected noise.

Noise cancellation (inverse sound wave) technology can also be used.

Remember: Most Hazards are Detected by Sight and Sound

The remaining three senses discussed should not be relied upon for hazard identification because:

- You have to be in contact with the hazard in order to sense it
- Human senses may be dulled or deadened by prior exposure to the hazard (hydrogen sulfide) or simply incapable of sensing the hazard (nitrogen asphyxiation)

- The extremely low levels at which certain hazards are lethal may preclude detection by senses – established technology, such as gas detectors, must be relied on!

3.2.3   Smell

The sense of smell can alert us to a hazard that is not detected by sight or sound. Colorless gases, for example, can often be detected by smell. While the sensors in the nose are very good at detecting the presence of an odor, they are not good at making absolute identifications of odors. In addition, some chemicals cannot be detected by the human nose at all. Research indicates that untrained workers can identify 15 to 32 common odors (Ref. 3-1). Trained workers can identify up to 60 odors. In addition, we are not good at detecting more than 3 or 4 intensity levels of an odor.

In an industrial setting, odors can provide an indication of hazards, as illustrated in Table 3-6.

Table 3-6.   Examples of odor hazard indicators

Odor Hazard Indicators	Possible Cause	Hazard Effects
Rotten eggs	Sour gas or hydrogen sulfide release	Lethal in low concentrations
Burning rubber	Friction or slippage	Overheating and possible mechanical failure
Ozone	Electrical arcing	Arcing, explosion, electrical fire, toxic gas
Smoke	Fire	Fire, explosion, toxic gas
Sweet aromatic	Hydrocarbon release, spill, vapor cloud	Fire, explosion, toxic gas
Diesel fumes	Building ingress through ventilation system	Carbon monoxide poisoning
Hot paint/hot plastic	Overheated equipment	Loss of containment Equipment failure
Acrid bleach	Chlorine or fluorine leak	Toxic gas Skin burns

The deliberate use of smell as a warning is not common, but some examples are:

- Mercaptan odorant in natural gas
- Wintergreen odorant in the CO_2 of fire suppression systems
- Stench odor in the HVAC systems of mines to alert miners to evacuate (In this case, odor is a faster and more reliable alerting mechanism than sound or vision.)

Facilities should examine their potential hazards to identify areas where the insertion of a recognizable odor can alert workers to hazards that would otherwise not be detected.

Finally, if the smell goes away, don't assume that the hazard is gone. Some toxic gases, like H_2S, desensitize your sense of smell and you can no longer detect it.

> **Some materials are extremely toxic and can be lethal
> at concentrations lower than their ability to be detected by smell.**

To prevent releases of these hazardous substances, instrumentation and engineering safeguards must be emphasized.

3.2.4   Touch

Sensors in the skin provide a wide range of inputs to the human. These include:

- Pressure sensors (e.g., touch)
- Pain sensors
- Temperature sensors

Some sensors serve a dual role, e.g., mechanical pressure and thermal. Physical contact can be used for two major purposes:

- To recognize potential hazards
- To determine whether a process or a machine is operating properly

As previously discussed, touching an object is not an ideal method for hazard identification. Touching an object can reveal its temperature (hot or cold), vibration, sharpness, and surface roughness; however touching an object is also a potential hazard to personnel. Table 3-7 illustrates some of the hazard indicators associated with various skin sensations.

Table 3-7.   Examples of skin sensation hazard indicators

Skin Sensation Hazard Indicators	Possible Cause	Hazard Effects
Cold pipe resulting in brittle line fracture	Loss of containment	Fire and explosion Worker injury
Hot surface	Loss of containment Insulation removed	Fire and explosion Worker injury
Line/equipment vibration resulting in equipment failure	Loss of containment	Fire and explosion Toxic release Worker injury Property damage
Rough surface (not visible)	Corrosion	Loss of containment Fire and explosion Toxic release Worker injury Property damage
Wet	Loss of containment	Fire and explosion Toxic release Worker injury Property damage
Sharp edges	Improper maintenance	Worker injury

Tactile displays have a history of use by the disabled. Hearing impaired individuals use tactile displays to localize sounds or to receive messages. Visually impaired individuals use the Braille coding system to read with, tactual maps for orientation, and vibrating canes for guidance. Shape coding can be used to identify controls and their location.

Highly skilled operators can use touch sometimes to detect hazards or abnormal operations. For example, a pulp and paper operator can tell by touch if there is the right amount of moisture in a roll of paper.

3.2.5   Taste

The sense of taste is closely aligned to that of smell. Taste provides the ability to distinguish chemical composition of two different substances. However, given the extreme risks associated with oral ingestion of some substances, taste is not employed as a hazard sensing method in the oil, gas and chemical sectors.

In the food and pharmaceutical industries, taste is sometimes used to determine whether the product is desirable. However, such taste testing only occurs after the absence of the hazard has been established by other means.

3.3   RELATIONSHIP BETWEEN SENSES AND HIGHER ORDER PROCESSES

Studies have shown (Ref. 3-9) that humans act as "single channel" processors of information.   Information from the senses is continually impinging on higher order processes and temporarily held in short-term memory. The input information is selectively filtered before it is processed according to a limited set of rules or principles about the state of the stimuli and that of the human.   The rules are paraphrased as follows:

- The human information process system can only process one sensory stimulus at a time
- A filter selects inputs from the senses for processing.
- The probability of any one of the five basic senses being selected increases based on the following properties:
    - Physical intensity (e.g., a loud sound or a bright flash will immediately attract attention)
    - Time since a similar event was last processed (the shorter the time, the higher the probability)
    - Sounds as opposed to visual stimuli
    - High frequency sounds as opposed to low frequency

The above principles have huge implications on hazard detection and identification. They imply that a worker who is mentally engrossed in a task may not detect a random hazard unless the hazard is:

- Physically intense, or
- The worker is expecting it to occur, or
- The worker has experienced the hazard recently

The principles also imply that the effect of distracting stimuli will likely be greatest when a worker is fatigued (e.g., at the end of a work shift).

The remainder of this section provides additional structure to the above principles. The goal of this section is to understand the problem of hazard detection and recognition.

Attention is a key to hazard recognition. It has been suggested (Ref. 3-10) that "... the term 'mind' points to a variety of functions of the brain - thinking, feeling, intending, perceiving, judging and so on - whereas the term 'mindfulness' or 'attention' points to a characteristic way in which any of these functions can move to center stage (or can move other functions off stage at any given moment)."

Attention is one of the three main limits on information processing along with memory and response time (Ref. 3-11). Attention is necessary to hold information in working memory and to move information to long-term memory. It is a vital component underlying decision-making and is closely aligned with perceptual processing.

Memory is the other aspect of information processing that is important to understanding the role of higher mental processes of hazard detection and identification. Memory can be divided into three types:

- Sensory Memory: Retains information only for as long as it is necessary to pass it on (usually less than a second).
- Working Memory: Receives information from sensory memory and retains it for 10-20 seconds, unless kept active through rehearsal.
- The research in working memory has practical implications on the way worker tasks should be designed:
    - Workers should not be given instruction verbally.
    - If workers must remember information for a short period of time, it should not be acoustically confusing.
    - Increase the time between successive messages, so they won't interfere with each other.
    - Group information into chunks and provide it in a logical sequence.
    - Try not to distract the worker from the message.

- <u>Long-term Memory</u>: Receives information from working memory and retains it for an indefinite period. Long-term memory is the memory of our lives. Information enters long-term memory through working memory. The ability of an item to be remembered long-term is a function of how often it is repeated, where repeating may imply:
  - Rote repetition - repeating the items over and over
  - Repeated exposure - remembering a route because you drove it every day for years

People can often recall stories and images of places and events that happened years ago. However, recall may not always be accurate. And, since workers use long-term memory to make decisions, the inaccurate retrieval of information can cause incorrect decisions which can have devastating effects.

INCIDENT HIGHLIGHT: PIPELINE EXPLOSION, 2002

- **Major gas leak from oilfield pipeline**

- **Vapor cloud explosion and fire**

- **4 fatalities**

- **19 injuries**

Your ability to detect leaks through hazard identification and monitoring can prevent catastrophic incidents.

3.4   INFLUENCE OF HUMAN CAPABILITIES AND LIMITATIONS ON HAZARD IDENTIFICATION

At this point, it is appropriate to summarize the information presented thus far into a set of principles and recommendations that can be used for hazard identification.

3.4.1   Visual Detection

If the hazard is visual (e.g., a leak in a pipeline), it must be detectable by the worker.

- Enough light must be provided to detect the hazard. Therefore, all potentially hazardous locations in the plant should be well lit.
- Contrast should be maximized between the potential hazard and the visual background. Let's assume, for example, that we are concerned about visually detecting a leak of bromine during tank truck unloading in a loading rack such as that shown in Figure 3-21. We know that bromine is a brown color and that it is heavier than air. So, we would expect the gas to fall over the side of the truck and gather on the pad of the loading rack.

Figure 3-21.   Bromine truck unloading operation

If we wanted to maximize the probability that the worker would be able to visually detect the gas, what color would we paint the truck and floor of the loading rack? *Hint: Use Table 3-1 to determine acceptable colors for the truck and floor.*

Figure 3-22 provides contrast colors for bromine (brown) against basic background colors such as black, gray light brown and white. In your opinion, what is the best background color to use in this situation?

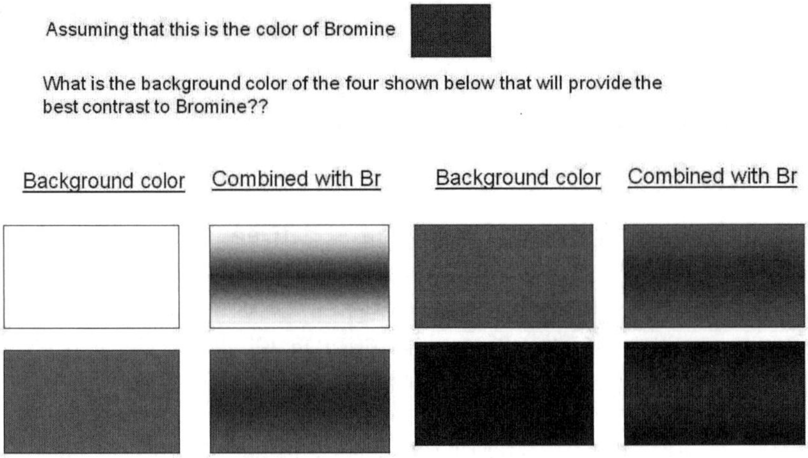

Assuming that this is the color of Bromine

What is the background color of the four shown below that will provide the best contrast to Bromine??

Background color   Combined with Br   Background color   Combined with Br

**Figure 3-22. Illustrating contrasts between bromine gas
and different background colors *(See color insert)***

- Hazards should not be present during the transition from light to dark adaption. If so, dark adaptation should be maintained by the use of red ambient illumination.
- Avoid excessive differences in brightness in the same visual field. For example, a control room operator should not have to look at bright ceiling lights located beside a visual display.
- Potential hazards should be located inside the worker's central field of vision. This is especially important when field workers are wearing bi-focal lenses.
- Potential hazards that are located outside of the central field of vision must be large and/or bright and/or move to attract attention.

- Colors should conform to work group expectations (e.g., red often means ON (energized) to electrical technicians and OFF (not running) to process technicians. Redundant coding should accompany color coding (e.g., shape, size, labels, etc.). Color deficient workers should not be used in color sensitive jobs (e.g., electricians).
- Potential hazards must be large enough to detect.
- Potential hazards must 'stand-out' from non-hazards.

3.4.2   Sound Detection

If the hazard can be heard:

- Warning sounds must be attention-getting
- Warning sounds should contain low frequencies so they 'carry' through structures
- Warning sounds should be unique. Wailing sirens of different patterns and frequency can convey different warning messages assuming that those heating it have been trained on what the warning means.
- Noise should not mask speech, warning signals or sounds relevant to job performance
- Loud noises or impulse noises (e.g., hammering) should be avoided since they may affect the concentration of workers
- Avoid the sudden onset of loud noises that can startle workers

3.4.3   Odor Detection

If the hazard can be detected by smell:

- Ensure that areas in which odors might be detected:
    - Are equipped with manual alarm call (MAC) points
    - Have easy access to emergency escape paths
- Where leaks in hazard streams might not be naturally detected, add an odorant to improve detection
- Train workers to detect odors that could potentially be present

> **Remember: Some toxic hazards can be lethal before they can be detected through smell.**

3.4.4   Touch Detection

If the hazard can be detected by touch:

- Identify areas and equipment where potential hazards could be present and detected by touch
- Ensure that the worker has easy access to areas and equipment where touch can reveal a potential hazard (e.g., rough or wet surface)
- Train workers to recognize potential hazards by touch
- Ensure that areas and equipment with which touch might be effective to detect hazards:
  - Are equipped with manual alarm calls
  - Have easy access to emergency escape paths

3.4.5   Hazard Recognition

To assist in the hazard recognition process, remember the following points:
- Attention is drawn to signals (hazards) that are:
  - Unique
  - Physically intense
  - Expected
  - Physically distinct
- Do not share similar mental resources between two or more concurrent tasks, e.g., cell phone use and vehicle tracking require the same mental abilities and will interfere with each other
- Do not share the same senses when performing concurrent activities
- Do not perform concurrent tasks that each require moving, positioning, orienting
- Do not perform concurrent tasks that share talking
- Do not require workers to maintain attention for long periods of time to signals that don't often occur and are hard to detect

- Minimize the need for short-term memory load. (e.g., chunk information, develop logical associations between activities, provide job aids or procedures)
- Minimize distractions while workers are performing tasks
- Simplify the information processing demands of tasks (e.g., automate calculations as much as possible)
- Items that are similar may be confused (e.g., the sounds 'e' and 'v' or the letters E and F)
- Items that can be visualized (e.g., a pump) are remembered better than abstract words
- The longer an item is rehearsed in short-term memory, the better it is retained in long-term memory:
  - Combined visual and verbal (out loud) rehearsal is better than visual or verbal alone
  - Rehearsal that relates material in new ways produces better recognition than rote rehearsal
- Linking items to memorable events or people, etc., improve the accessibility of those items in memory (e.g., creating a word to represent a process or procedure is a strategy that organizes materials in memorable ways)
- Deep processing of items can increase retention (Ref. 3-15). 'Deeper' processing refers to an analysis of meaning, inference and implications, contrasting 'shallow' processing such as surface form, color, loudness, brightness, etc.
- Information memorized under stress may not be retained as well as information memorized without stress
- Recall is better for lists of words that belong in the same categories
- Organizational techniques can improve long-term memory
- Items that are associated with retrieval cues are remembered better when the cues are presented than when they are not. Vehicle names (Chevy Nova) that are associated with manufacturers (General Motors) are better remembered when manufacturers are used as cues to recall.
- External memory aids are used to help people remember to do things rather than remember what to do. Procedures, for example, are aids to recall by reminding workers what they need to do rather than telling them how to do it.

- Items that are organized in a logical sequence are better remembered than those that are not organized.

3.5   WHAT CAUSES HAZARDS?

Process hazards are created by the intrinsic (i.e., inherent) properties of the materials used in the process and the conditions under which they are stored, used, processed or otherwise handled.   For example, the flammability hazard of ethylene is intrinsic to the material, and as such, use of ethylene can lead to process hazards such as fire and explosion.

The following sections provide examples of hazards that result from failures of the safety systems, failure of others that jeopardize the health and safety of the worker or results in a loss to the company, or failures of the worker.

3.5.1   Example 1

In the fall of 1998, a chemical worker was badly burned in an explosion and fire at an ethylene reactor. The reactor was being re-started after a shutdown. One of the restart operations was to purge three catalyst lines of old catalyst and refill them with fresh catalyst. The reactor is illustrated in Figure 3-23.

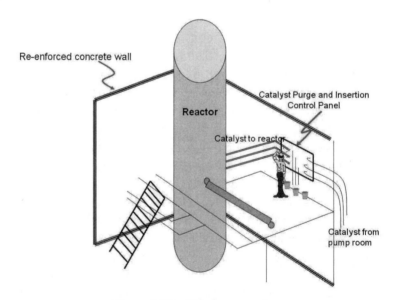

Figure 3-23. Ethylene reactor

The worker operated the control panel from a platform accessed by a ladder from grade. The worker had to duck under the catalyst lines entering the reactor to access the catalyst purge and insertion control panel. The purging process required that each catalyst line - top, middle and bottom - be purged, one line at a time. For any one line, the process began by closing the reactor injection valve (Figure 3-24), opening the catalyst valve and purge valve, and starting the catalyst pump to purge the old catalyst into a simple metal pail that sat on the platform grating (Figure 3-25). The valves shown in Figure 3-24 were multi-turn and were milled from a block of metal. As a result, the valves did not visually indicate fully open or fully closed and required the worker to turn them fully clockwise or counterclockwise to check their status.

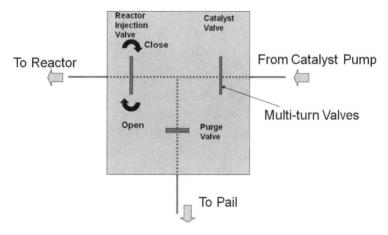

Figure 3-24. Catalyst valve block

As it turned out, in this case, the reactor injection valve on one line was not fully closed allowing ethylene under pressure to push back and drain into the pail. It was thought that a static spark ignited the ethylene in the pail, severely burning the worker standing at the panel. The worker was unable to call for help or escape owing to the obstructions shown in Figure 3-23.

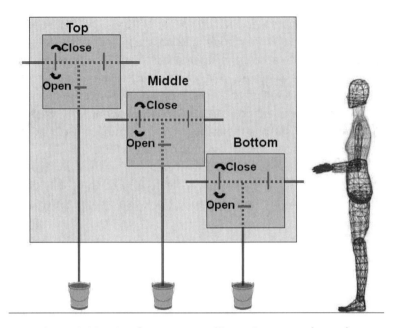

Figure 3-25.   Catalyst purge and insertion control panel

The worker made an error by not checking to see if the reactor injection valve was fully closed before opening the purge valve and by not grounding the metal pail. But is he the root cause of this life threatening incident? Clearly no, since the design of the purge system did not take into account several issues:

- The valves did not visually indicate position
- The design did not consider worker egress
- No safeguards were designed into the system to prevent back pressures, out-of-sequence valve operation, lack of grounding, etc.
- Escape routes from the platform were unacceptable

In fact, the operation could have been conducted from a remote location using readily available control valves and sensors. This accident did not have to happen.

3.5.2   Example 2

In 2004, a sudden release of flammable liquid in an HF Alkylation Unit caused a fire and explosion that injured six workers and caused significant property damage (Ref. 3-16).   An alkylate pump was not properly isolated prior to maintenance. When the pump was dismantled, alkylate sprayed out at 150 psi covering the maintenance crew and then caught on fire.

Prior to beginning pump maintenance, both the operations and maintenance crew thought that the suction and discharge valves were closed and locked out. Figure 3-26 illustrates the pump suction valve. The handle on the valve indicates that the valve is closed, but a closer look at the valve position indicator shows that the valve is really open. The valve handle is installed incorrectly.

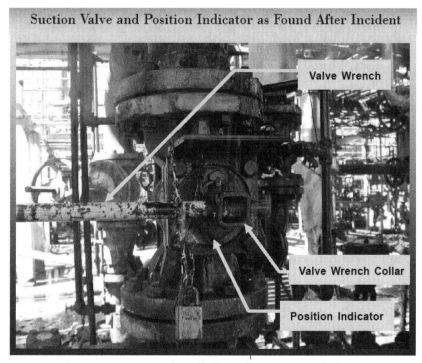

**Figure 3-26.   Pump handle is 90 degrees out-of-phase
with pump position indicator**

*The narrative and photos to follow can be found on the US Chemical Safety Board
website.*

After 'isolating' the valves, the vent hose to flare was removed (Figure 3-27). A small amount of alkylate was released but soon stopped. The mechanics assumed that the pressure was released and the pump was free of product. As it turns out, scale had blocked the flare vent outlet. The pump was still pressurized. When the mechanics released the flange bolts, alkylate sprayed out.

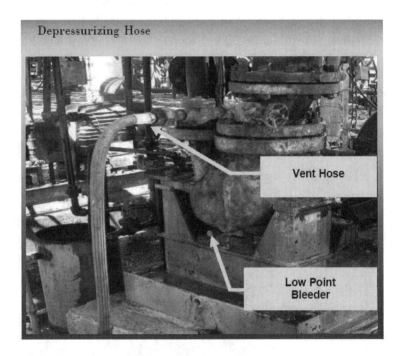

Figure 3-27. Flare vent hose did not indicate pressure in the pump

This case provided a number of lessons for learning:

- Workers expect that a ¼-turn plug valve is closed when the valve handle is perpendicular to the direction of flow
- Workers did not check the position indicator on the valve stem to determine open/closed status
- Prior to the incident, the valve mechanism was changed from a gear and wheel actuator to a handle - the handle was installed incorrectly

- There are several management systems that should have detected or prevented the incorrect installation: Management of Change (MOC), Process Hazard Analysis (PHA), and Pre-Startup Safety Review (PSSR)

Clearly, this hazard was the result of both the failure of other workers to recognize the problem and the failure of the safety management system that allowed the hazard to exist in the first place.

3.5.3   Example 3

One of the most common operating mistakes made in a chemical plant is "cross-over" errors. In this situation, a product is mistakenly routed through a manifold (Figure 3-28) to the wrong tank, thus contaminating the existing product in the tank.

Figure 3-28.   Manifold for directing product into and out of storage tanks

The mistake can be very costly if the contaminated product is then used for an application that causes equipment failure. (Several years ago, in Australia, contaminated aviation fuel allegedly caused the engines on several general aviation aircraft to fail). Cross-over errors happen for at least three reasons:

- Instructions are given verbally without a hard-copy follow up. This generally happens in situations where manifolds are used to direct product to a particular tank. Without hard-copy backup, the original instructions are forgotten or are recalled incorrectly.
- 'Absent-minded' errors are made. These generally occur in situations where the task is routine. Routine tasks are 'automated' and not continually attended to. Attention can easily be directed to another sensory stimulus making the routing task prone to error.
- Poor identification of valves and piping. Labeling or tags that are missing or not legible can lead to inaccurate procedures and valves being placed in incorrect positions, thus contributing to product cross-contamination.

3.6   REFERENCES

3-1.   Sanders, M.S., and McCormick, E.J. "Human Factors in Engineering and Design: Seventh Edition." McGraw-Hill, New York, NY. 1993.

3-2.   Attwood, D.A. "The Office Relocation Sourcebook." John Wiley and Sons, New York, NY. 1996.

3-3.   Attwood, D.A., Deeb, J.M., and Danz-Reece, M.E. "Ergonomic Solutions for the Process Industries." Gulf Publishing (Elsevier), Burlington, MA. 2004.

3-4.   NSI. "American National Standard for Product Safety Signs and Labels." American National Standards Institute, Standard No. Z535.4-2002. 2002.

3-5.   International Standards Organization. "Petroleum and Natural Gas Industries: Offshore Production Installations – Guidelines on Tools and Techniques for Hazard Identification and Risk Assessment." International Standards Organization, Standard No. 17776. 2000.

3-6.   NIOSH. "Criteria for a Recommended Standard: Occupational Noise Exposure". National Institute for Occupational Safety and Health, Publication No. 98-126. 1998.

3-7.   Attwood, D.A. "Environmental Factors." Chapter 15 in D. Crowl (ed) Human Factors Methods for Improving Performance in the Process Industries. Sponsored by the Center for Chemical Process Safety, John Wiley and Sons (publishers), Hoboken, NJ. 2007.

3-8.   Eastman-Kodak. "Ergonomic Design for People at Work: Volume 1." Van Nostrand Reinhold, New York, NY. 1983.

3-9.   Broadbent, D. "Perception and Communications." Pergamon Press, London. 1958.

3-10.   LaBerge, D. "Attentional Processing: The Brain's Art of Mindfulness." Harvard University Press, Cambridge, MA. 1995.

3-11.   Wickens, C.D., and McCarley, J.S. "Applied Attention Theory." CRC Press, New York, NY. 2008.

3-12.   Miller, G.A. "The Magical Number Seven, Plus or Minus Two: Some Limits on our Capacity for Processing Information." Psych Review, 63, 81-97. 1956.

3-13.   Chapanis, A., and Moulden, J.V. "Short-Term Memory for Numbers." Human Factors, 32, 123-137. 1990.

3-14.   Peterson, L.R., and Peterson, M.J. "Short-Term Retention of Individual Verbal Items." J. Exp. Psych., 58, 193-198. 1959.

3-15.   Craik, F.I.M., and Lockhart, R.S. "Levels of Processing: A Framework for Memory Research." J. Verbal Learning and Verbal Behavior, 11, 671-684. 1972.

3-16.   Giant Industries. "Case Study Oil Refinery Fire and Explosion." U.S. Chemical, Safety Hazard and Identification Board, Case Study Number 2004-08-I-NM, October. 2005.

4

TYPES OF HAZARDS

A comprehensive understanding of the types of hazards in your workplace is essential to the hazard management process (Figure 4-1). Hazards range in complexity from the fundamental, visually identifiable, to the complex requiring not only the senses, but intentional focus and analysis as well.

> Fundamental hazards can often be identified by walking through the process unit specifically looking for the existence of potential hazards. Every walkthrough of the process unit is an opportunity to identify potential hazards.

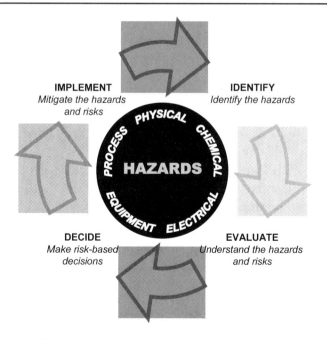

Figure 4-1.   Hazard management process

Beyond the walkthrough, there are opportunities to be more deliberate and focus on specific areas of the process unit, such as:

- Scheduling and coordinating maintenance work
- Running permit approvals
- Taking field samples and readings
- Conducting focused, checklist driven inspections

Being in the process unit is not the only opportunity to identify hazards. Hazards can be identified in many ways, including:

- When operating the unit from the control boards or screens, be aware of:
  - Status of instrumentation and alarms
  - Confusing/conflicting alarms and alarm overload
  - Control response changes
  - Long term or permanent changes in processes or conditions that warrant re-prioritizing of alarms
  - Special conditions that may affect alarm states, such as start-ups or shutdowns
  - Addition or modification of interlock systems
- When reviewing or writing operating procedures, be aware of:
  - Bypassing or disabling interlocks or automatic action systems
  - Tests and inspections not being conducted according to schedule
  - Operating procedures not representing actual practice
  - Ambiguous or unclear procedural steps that could lead to misinterpretation or confusion

---

The key is paying attention to the possibility that hazards exist.

The previous chapter discussed how we - as human beings - recognize hazards using our various sensory capabilities. In this chapter, various types of hazards are discussed and illustrated as examples of hazards that can be identified through our human capabilities.

These hazards can be categorized as:

- Workers hazards that have minimum potential for process implications
- Technical, mechanical or operational hazards that can lead to loss of containment of hazardous materials or other serious process hazards

Obviously, these are not mutually exclusive. Some workers hazards could lead to loss of containment and loss of containment can certainly present hazards to workers. This book emphasizes the identification and recognition of process hazards.

> **The examples in this book illustrate that many types of hazards may exist concurrently.**
> **An improperly placed container may fall and rupture releasing hazardous material to atmosphere. A worker could be injured by the falling container or by the material that is released.**

4.1   EXPLOSION HAZARDS

4.1.1   Reactive Explosion Hazards

Reactive explosion hazards are caused by the reaction of two or more materials. Large explosions can be caused by mixtures of reactive chemicals. Contamination leading to chemical explosion can occur in a number of ways. Manifolds and other multiple connection points (Figure 4-2) are sometimes the cause of reactive explosion hazards.

Figure 4-2.   Manifold - potential for cross contamination
of reactive chemicals

INCIDENT HIGHLIGHT: EXPLOSION, 2001

- Ammonium nitrate storage

- Explosion and fire

- Facility destroyed

- 31 fatalities

- 2,440 injuries

- Damage $800MM

Incompatibilities between chemicals and also with materials of construction (MOC) should be evaluated and documented, most often using a binary chemical/material interaction matrix. Binary interaction matrices can be used to help ensure safe operation of a plant for both intended reactions and unplanned reactions that may result from inadvertent mixing of chemicals. The binary chart usually indicates if the two chemicals or materials are reactive under the given conditions. Information for filling the matrix can be obtained from literature sources or through experiments.

The Chemical Reactivity Worksheet is a free program containing information about the reactivity of substances or mixtures of substances that is available from the National Oceanic and Atmospheric Administration (NOAA) website (http://response.restoration.noaa.gov/chemaids/react.html). The worksheet has a database of reactivity information for more than 6,000 common hazardous chemicals and can be downloaded free of charge.

Batch processes (Figure 4-3) offer a means of introducing errors in mixing reactant chemicals due to:

- Requirements for adding various process chemicals and other ingredients often at various stages of the batch (Many of these ingredient additions are performed manually.)
- Liquid ingredients are piped to manifolds at reactors with manually operated valves for charging to the reactor. Manifold valves inadvertently left open can result in cross contamination. Depending on the nature of the process chemicals, there can be severe reactions.
- Batch reactors often require cleaning between batches, particularly when they are used for making different products. Cleaning fluids that are incompatible with the process chemicals (like water or solvent) may remain trapped in the system and cause inadvertent mixing and severe reactions.

Figure 4-3.   Increased probability of inadvertent mixing or overcharging of chemicals in batch processes

4.1.2   Flammable Explosion Hazards

Release and ignition of flammables into areas of high congestion or confinement can produce blast overpressures that can destroy process unit structures and control rooms or other buildings.   Other severe results include projectiles and flying debris, such as glass shards from building windows. Workers in these buildings are at risk for severe injury or fatality.

An example of a congested area may be an outdoor petrochemical processing unit with high or medium equipment density, and an example of a confined area is a flammable process unit inside a building. When the flammable vapor cloud is ignited in one of these congested or confined areas, the pressure wave is obstructed in one or more directions in its movement away from the point of ignition.   These obstructions cause increases in strength of the pressure wave.

INCIDENT HIGHLIGHT: FIREBOX EXPLOSION, 2000

- Firebox explosion

- 2 fatalities

- 1 injury

- Startup problems

- Safety features bypassed

- Inadequate MOC

In colder climates, excess materials and equipment may be staged around process plants to provide a windbreak for workers. This can create confinement and may entrap flammable vapors which can contribute to a vapor cloud explosion (Figure 4-4).

Figure 4-4.   Windbreaks can create confinement

By contrast, if the release of flammables is into an open area where the pressure wave from the ignition point can move freely in all directions, a significant overpressure wave may not develop. Significant overpressure waves have catastrophic consequences (Figure 4-5).

Figure 4-5.   Aftermath of a flammable release and explosion
(See color insert)

INCIDENT HIGHLIGHT: BOILER EXPLOSION, 2004

- Steam boiler explosion adjacent to LNG containment

- Large release of LNG and subsequent explosions

- 23 fatalities

- 74 injuries

- Facility destroyed

- Damage $800MM

4.1.3   Physical Explosion Hazards

Physical explosion hazards include sudden and violent releases of a large amount of gas/energy due to a significant pressure difference. Examples of physical explosion hazards include a rupture of a pipeline, a boiler, compressed gas cylinder, or pressure vessel (Figure 4-6). The size of the energy release is related to initial pressure and system volume.

Figure 4-6.   Ruptured pressure vessel *(See color insert)*

Pipelines may fail in service as a result of improper assembly, corrosion or overpressure. When this occurs, the pipeline may rupture catastrophically, as shown in Figure 4-7.

Figure 4-7.   Catastrophic rupture of a natural gas pipeline

Overpressure hazards can also be caused by weakened flanges or covers due to missing bolts (Figure 4-8). Plant workers should periodically check bolted connections to determine if they have been assembled with their specified number of bolts and for tightness, particularly following turnaround and maintenance. Some companies have requirements to ensure that two threads penetrate through the nut.

Figure 4-8.   Missing bolts in flanges can weaken the joint

Overpressure hazards can occur when pressure relief devices (Figure 4-9) fail due to:

- Blocked or clogged weep holes in relief device piping. Blocked weep holes in relief device discharge piping can create the hazard of accumulation of water which may prevent the relief device from properly operating.   Weep holes should be checked visually for indication of blockage.

Figure 4-9.   Rupture disc protecting a pressure relief valve with pressure gauge monitoring of the interstitial space between them
(See color insert)

- Damage from exposure to process streams. Pressure gauges connected to interstitial spaces between relief valve and upstream rupture disc provide an alert that the relief valve has been exposed to the process stream. These pressure gauges are installed to protect the relief valve internals in corrosive, polymerizing services. A pressure reading indicates that the rupture disc has failed and material has possibly leaked into the space. This can lead to corrosion, causing the relief valve not to function or not to function as designed.
- Corrosion due to process gas exposure
- Plugging of inlets or vent lines (Figure 4-10)
- Inadequate design, installation and maintenance

Figure 4-10. Bird's nest in outlet of safety relief valve *(See color insert)*

4.2   CHEMICAL HAZARDS

4.2.1   Toxic Chemical Hazards

Toxic chemicals result in illness, disease, or death by interfering with the body's biological processes. Chemicals may be inhaled, absorbed, ingested, or injected, and if their rate of entry exceeds the body's rate of elimination, the chemicals become toxic. Large spills of toxic chemicals (Figure 4-11) can put entire communities at risk of injury, illness or fatality.

Figure 4-11.   Release of toxic chemicals

Many chemicals in a process plant are toxic to humans. The toxic effects typically vary with contact time and type of exposure (e.g., skin contact versus ingestion or inhalation). This section examines the many ways in which plant workers can come into contact with toxic chemicals. In all instances, PPE should be provided and worn to help prevent possible exposures.

4.2.1.1   Process Sampling

It is important that proper sampling procedures are followed and the sampling station is well designed to avoid personnel exposure to toxic materials. Many sample stations:

- Are too congested for convenient access (Figure 4-12)
- Have the sample valve and sample container too far apart, making it difficult for the operator to control sample flow while checking the level in the container
- Locate the sample box in awkward locations (Figure 4-13)

Figure 4-12.   Limited clearance for sampling

One hazard often leads to another.

What types of hazard could be introduced when trying to use this drain valve?

How would you resolve this hazard?

Figure 4-13.   Sample container in a hard-to-access location

4.2.1.2   Indoor Locations

When indoor locations are used for shelter-in-place they should be designed and operated to ensure that toxic or flammable gases do not enter during an emergency:

- Shelter-in-place rooms, including control rooms, should have HVAC shutdown switches, which should be clearly identified and located for ready access. Failure to promptly shutdown HVAC presents potential exposure hazards to those who have or may seek shelter in the control building.
- Control buildings designated as shelter-in-place should ensure that the windows, doorways and openings through walls - conduit, ventilation ducts, etc. - are sealed and the sealing material completely fills the open spaces. If not, the control building is potentially vulnerable to ingress of hazardous vapors or gases (Figure 4-14).

Figure 4-14. Ensure vents into buildings are not located
near process vents

4.2.1.3   Preparing Equipment for Maintenance

In preparation for opening process piping with a safe work permit, such as lockout/tagout (LOTO), there is a need to review isolation, bleed down and drainage, as well as proper PPE. It is also important to review procedures for opening the first flange to prevent workers being exposed to direct contact with the last material in the system. Finally, drain lines should be accessible and positioned to appropriate discharge locations (Figure 4-15).

How many hazards can you identify?

Is the location of the drain appropriate?

Is the PPE appropriate for the job task?

Has the equipment been properly isolated using LOTO procedure?

Have all worker groups communicated adequately?

What are your company's policies regarding draining equipment for maintenance?

Figure 4-15.   Draining equipment for maintenance

4.2.1.4   Confined Spaces

Confined space entry jobs must be checked to determine if hazards have been created by lack of oxygen, the space not being completely isolated, chemicals and energy not totally removed, insufficient purging, proper PPE not being worn, and the attendant not actively monitoring the work inside the space. Figure 4-16 illustrates access horizontally through a manway.

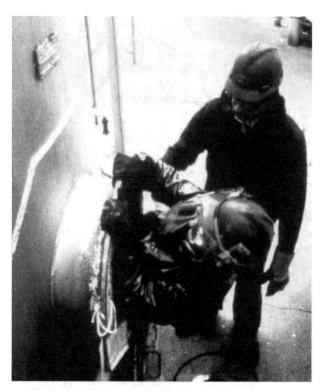

Figure 4-16. Confined space entry

4.2.1.5   Outdoor Locations

Workers in outdoor locations can be exposed to toxic material releases. Many facilities are equipped with toxic detectors, like hydrogen sulfide detectors, to warn personnel of potential releases and exposure. Employees are trained to evacuate the process unit, traveling upwind to a designated shelter-in-place location or designated evacuation area.

It is important to be sensitive to significant changes in odors. Almost by definition, odors of process chemicals are evidence of some loss of containment. It likely will require instruments to locate the source rather than allowing workers to locate and detect the source by smell.

4.2.2   Fire Hazards

Fire hazards involve a chemical that, when exposed to an ignition source, results in combustion. Large fires (Figure 4-17) can create toxic gases that can put communities and first responders at risk of illness and injury.

Figure 4-17.   Large chemical fires have potential for injury and illness
(See color insert)

When there is loss of containment of flammable materials, there is also a possibility that the material will contact an ignition source and fire will occur. To mitigate this risk, materials must be contained and ignition sources must be eliminated or controlled in areas where containment might be breached.

4.2.2.1   Process Hoses

Hoses in process service are vulnerable to failure due to physical damage and integrity of connections (Figure 4-18). Hose failure can create significant loss of containment or lead to personnel injury (due to hose whip) and must be checked periodically for damage and proper connections. Replacement hoses must be reviewed and inspected for appropriateness for the properties of materials and process conditions for which they will be used.

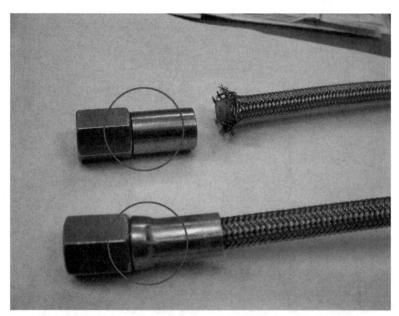

Figure 4-18.   Top hose connection defect received from manufacturer and burst in service

4.2.2.2   Open-Ended Valves

Check for open-ended valves in hazardous material services (Figure 4-19), particularly quick open type valves. These type valves can be caught in clothing or inadvertently moved by workers working around them. This can cause release of hazardous material and workers are often nearby. Such valves, especially indoors, should always be plugged or capped.

**Figure 4-19.   Open-ended valves in hazardous service can
be a source of chemical leaks**

4.2.2.3   Isolated Relief Devices

If one or more of the valves on the inlet or outlet to a relief device is closed, the relief device will not protect equipment from over pressure, leading to a significant fire, explosion and personnel injury hazard. Valves that are chained, car-sealed or locked open (Figure 4-20) must be verified and documented to ensure integrity of protective devices, such as relief valves and critical instruments.

**Figure 4-20.   Car-seal and chain for isolation valve
beneath relief valve**

4.2.2.4   Sight Glasses

Sight glasses and glass level gauges have been sources of catastrophic failures with release of toxic, corrosive or flammable materials. These glass units are typically "weak" points in the integrity of a system. Armored glass level gauges and specially designed vessel surface-mounted sight glasses have been shown to reduce the risk of failure. Workers should check to determine if a level gauge or sight glass is functioning as intended:

- Can the material that the level gauge or sight glass is indicating be readily seen?
- Is the location of the sight glass or level gauge at increased likelihood of being damaged by maintenance or other work activity?

If sight glasses or glass level gauges are not functioning as intended or are at increased risk of being damaged, they should be removed at the next down time opportunity. Alternatives, such as instrumentation are available for measuring level.

4.2.2.5   Ignition Sources

There are many ignition sources in process facilities that must be controlled, including:

- Hot work – Hot work permits and programs control the hazards associated with welding, use of non-intrinsically safe devices and other activities in areas that have the potential for flammable and explosive materials. A well-managed hot work program is essential in controlling ignition sources.
- Static – Bonding and grounding are necessary to prevent static discharges and potential ignition sources. See Section 4.3.4 for more details.
- Smoking – Smoking should only be permitted in designated areas to prevent the hazards associated with open flame in process areas where flammable or explosive materials may be present.
- Motor vehicles – Motor vehicles in process areas, such as cars, trucks, fork-lifts, or pallet jacks, represent the hazard of uncontrolled ignition sources in the event of a flammables release. These vehicles could be powered by either internal combustion (gasoline, propane) or non-classified electric motors. Vehicle entry into process areas should be controlled.

4.2.2.6   Critical Instruments and Equipment

Hardware that is necessary to avoid loss of containment leading to a major accident scenario must be regarded as critical. Critical instruments and equipment should be part of a rigorous preventive maintenance program. Critical instruments, switches, valves, piping and vessels must be clearly identified. Poorly labeled (Figure 4-21) or unlabeled equipment and facilities can create hazards.

Figure 4-21.   Poorly labeled valves can slow an operator's
response in an emergency

4.2.3   Corrosive Chemical Hazards

Corrosive chemicals are those that cause damage when they come into contact with skin, metal, or other materials. Acids and bases are examples of corrosive chemicals. Large spills of corrosive chemicals (Figure 4-22) such as concentrated sulfuric acid (oleum) or hydrochloric acid can create severe hazard potentials to workers in the vicinity and in the surrounding community.

Figure 4-22. A railcar release can impact an entire community

Common sources of release of corrosive chemicals are from leaking lines, valves and equipment (e.g., pumps), valves that are mistakenly opened, and storage vessels that are not compatible for the chemical. Additional causes include replacement equipment and parts that are not rated for the chemicals being contained. It is essential that when equipment and parts are being replaced, the replacement parts are verified to be correct - particularly in materials of construction and pressure rating.

Often, loss of containment is due to the failure to identify corrosive chemicals during job planning or permit completion. Or, loss of containment could be due to incorrect handling of chemicals due to improper procedures or procedures that are not followed correctly. It is essential that the safety management processes are in place in your plant and that they are followed correctly.

All workers should be trained to look for leaks, drips, and other indications of loss of containment. Leaks, drips, and, for some materials, smells can indicate the beginning of gasket or mechanical seal failure (Figure 4-23) and significant loss of containment. Be sure to check for these leaks both in open process structures, as well as in enclosed process or compressor buildings. Even small leaks of certain substances may present significant hazards to workers. Workers should check particularly around pumps and compressors and high pressure lines and equipment. However, workers should be aware that if a leak has partially or completely sealed itself with reactive byproducts, then flushing the area with a leak check fluid (e.g., soap solution) may rinse away the residue and significantly increase the leak rate.

A flange leak testing program can identify incipient leaks that can be addressed before they progress into large leaks. These will often be detected visually and sometimes by odor. If significant odor or a change in odor is identified, arrange for instrument support to identify the specific leak(s).

Figure 4-23. Leakage from a pump

4.3   ELECTRICAL HAZARDS

4.3.1   Shock/Short Circuit

Electrical hazards include contact with exposed conductors or a device that is incorrectly grounded, such as metal equipment coming into contact with power lines (Figure 4-24).

Figure 4-24. Equipment working near overhead power lines

These hazards are typically found in electrical substations or when working with overhead lines. In some cases grounding/bonding connections are broken or missing. Grounding or bonding connections serve to minimize the potential hazard of electrical charge accumulation and arcing. In the presence of flammable vapor-air mixtures, this can result in a fire. Grounding/bonding straps should be checked for corrosion, damage and continuity.

Improper use of electrical equipment can have catastrophic consequences, as illustrated by this incident.

INCIDENT – BALLAST TANK EXPLOSION

A bulk cargo ship undergoing maintenance to the ballast tanks sustained a violent explosion that killed 8 workers. The ship was temporarily anchored in a harbor and painting operations were underway in one of the hollow tanks mounted on the side of the vessel. The paint and solvent they were using were flammable. To provide breathing air and to prevent the workers from being overcome by fumes, an electric fan was placed over the entrance manway. This was periodically moved and adjusted to introduce fresh air and remove the stagnant air. An electrical cord was connected to the fan. Without warning, a violent explosion occurred within the tank and three workers were killed instantly. Several others were blown off the deck by the force of the explosion; some of their bodies were later recovered.

From this brief explanation, how many hazards could have contributed to this incident?

Another situation that can lead to electrical hazards is when lockout/tagout (LOTO) equipment (Figure 4-25) is not properly installed or LOTO procedures are not being followed. This can occur because:

- Job planning was not properly conducted
- Plant workers failed to identify and isolate all energy sources
- Lockout/tagout protections do not remain in place during the task

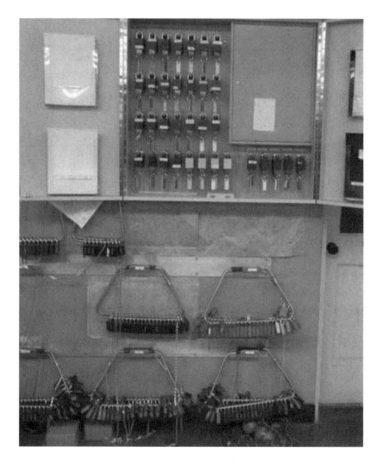

Figure 4-25.   One example of a lockout closet in a process plant

4.3.2   Fire

Electrical fire hazards include use of electrical power that results in electrical overheating or arcing to the point of combustion or ignition of flammables, or electrical component damage.

Overloading electrical connections is another concern. At times, the overload is deliberate because one electrical strip is plugged into another until the load on the outlet is excessive. In other cases, the overload is inadvertent. The combination of outlets and equipment on a single breaker is distributed so widely that an operator may not know that the lines are overloaded. If the breaker is damaged, fire can ensue. It is important for plant workers to understand the loads on a single breaker and to inspect the equipment to ensure that it is not damaged.

Sometimes, part of a process is energized, for example electrolytic processes or a refinery desalter. In such cases, proper electrical isolation of equipment (e.g., insulating flange) is very important.

4.3.3   Lightning Strikes

Electrical storms generate large discharges to other clouds or to the ground. Prominent objects tend to concentrate charges in the earth as the electrical storm approaches. The discharge of lightning is drawn to those charges. People in the open or near to prominent objects, particularly non-conductors such as trees, are at risk if a lightning strike occurs. The charge attempts to flow to the ground, but the paths to the ground are not sufficient. The charge then propels off and can seek nearby workers where secondary charges may have accumulated. When a storm approaches as predicted by the weather services or announced by the sound of thunder, the optimum hazard management approach is to stay indoors.

4.3.4   Static Electrical Discharge

Electrostatic discharge can develop by charge separation inherent in flowing liquids or solid particles. All objects (both conductive and non-conductive) have an inherent electric field due to electrons in the material. However, if one object is less conductive, electrons are unable to move fast enough and uneven distribution of electrons results in two opposite but equal charges. This unequal accumulation of charges can lead to static discharges (spark, brush, corona, etc.) that may serve as an ignition source.

Rubbing of wool, nylon, other synthetic fibers, and even flowing liquids can generate static electricity. The discharge (spark) to the ground results in the ignition of flammables. Grounding and bonding are necessary to prevent static electricity hazards (Figure 4-26).

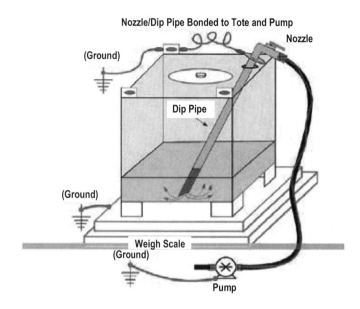

**Figure 4-26.   Recommended grounding for filling tote
with flammable liquid**

In 2007, a flammable liquid explosion and fire at a solvents distribution facility led to the evacuation of 6,000 residents. The fire destroyed the facility. A key lesson learned is to ensure that equipment, such as fill nozzles and hoses, is bonded and grounded and designed for flammable use (Ref. 4-1).

4.3.5   Loss of Power

Safety-critical equipment failure as a result of loss of power is a significant hazard. Loss of electrical power is a serious abnormal event in a plant. Loss of power is also an event that should be expected to occur and be properly planned for. If no emergency power is available, critical equipment, such as compressors, will shutdown. Work performed on electrical transmissions may lead to power interruptions (Figure 4-27).

Figure 4-27. Work on electrical transmission lines

4.4   EXCAVATION HAZARDS

Soil collapse in a trench or excavation as a result of improper or inadequate shoring presents a hazard to workers.

Workers working in excavations more than 5 feet deep must have the excavation protected against caving-in onto them (Figure 4-28). Even excavations less than 5 feet deep (Figure 4-29) must have some assessment of the potential for cave-in. Soil from the excavation must be placed at least 2 feet from the edge of the excavation.

What are the hazards in this picture?

What could happen if they are ignored?

What can you do about it?

Figure 4-28. Excavations must be designed to protect workers from cave-in

Figure 4-29. All excavations must be assessed for the potential of cave-ins

4.5   ASPHYXIATION HAZARDS

Nitrogen is an asphyxiation hazard.   It should not be used to power compressed air tools.   Users of compressed air tools should verify that workers have not attached their air hoses to nitrogen piping.   Standardize and use connections for nitrogen that cannot be misapplied to utility connections (Figure 4-30).

Figure 4-30.   Ensure that nitrogen cannot be inadvertently connected to any other utility connection

4.6   ELEVATION HAZARDS

Workers working at elevations without fall protection (Figure 4-31) are at risk of injury. It is essential to identify situations where workers are exposed to fall hazards by working at elevations without fixed or personal fall protection. Workers working outside of the boundaries of fixed fall protection must use harness and lanyard systems (Figure 4-32). Fall protection is an essential part of any job requiring elevated work.

Figure 4-31. Workers must not work outside of a safety cage without wearing fall arrest systems

Figure 4-32. Working at heights with personal fall protection

Warning: This risk of workers injury still exists even when wearing appropriate fall protection. Even though fall protection safeguards are being employed, a fall may require the worker to receive extensive medical attention.

It's time to break the cycle of incidents that occur in the process industry - and establishing a strong workplace hazard management program is an essential step in this process. (See Page 251)

What types of hazard controls would you apply to the above job tasks?

How can management commitment and employee ownership affect these controls?

(See Page 246)

Large chemical fires have potential for injury and illness (See Page 111)

Tools not being used for their designed purpose (See Page 142)

Offshore platform in hurricane conditions

(See Page 147)

Wind damage at polymer facility

(See Page 149)

Process facilities impacted by flooding (See Page 148)

Lifting load over people

(See Page 21)

From this distance, a person with a color deficiency would not be able to determine the alarm conditions on this panel

(See Page 56)

Assuming that this is the color of Bromine

What is the background color of the four shown below that will provide the best contrast to Bromine??

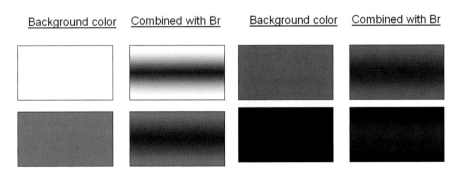

Background color | Combined with Br | Background color | Combined with Br

Illustrating contrasts between bromine gas and different background colors

(See Page 80)

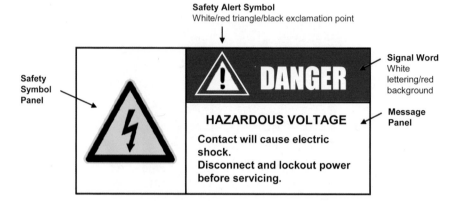

Safety Alert Symbol
White/red triangle/black exclamation point

Safety Symbol Panel

Signal Word
White lettering/red background

Message Panel

⚠ DANGER

HAZARDOUS VOLTAGE

Contact will cause electric shock.
Disconnect and lockout power before servicing.

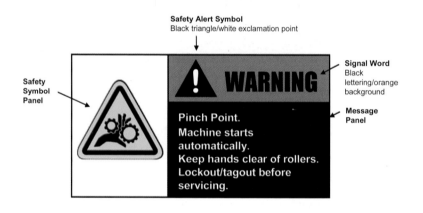

Safety Alert Symbol
Black triangle/white exclamation point

Safety Symbol Panel

Signal Word
Black lettering/orange background

Message Panel

⚠ WARNING

Pinch Point.
Machine starts automatically.
Keep hands clear of rollers.
Lockout/tagout before servicing.

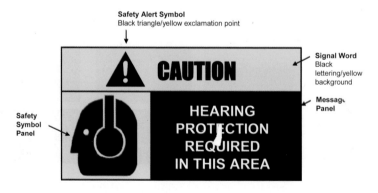

Safety Alert Symbol
Black triangle/yellow exclamation point

Signal Word
Black lettering/yellow background

Message Panel

⚠ CAUTION

Safety Symbol Panel

HEARING PROTECTION REQUIRED IN THIS AREA

(See Page 67)

Hazard identification methods can be used to identify hazards associated with facilities and job tasks, like those pictured above   (See Page 153)

Corrosion under insulation causes pipe failure and valve malfunctions or leak

(See Page 29)

Combustible structure around flammable gas system

(See Page 30)

Exposed electrical wiring

(See Page 20)

Manifold - potential for cross contamination of reactive chemicals (See Page 34)

Aftermath of a flammable release and explosion (See Page 100)

Ruptured pressure vessel (See Page 101)

Rupture disc protecting a pressure relief valve with pressure gauge monitoring of the interstitial space between them (See Page 103)

Bird's nest in outlet of safety relief valve (See Page 104)

Cleaning the exterior of process equipment (See Page 196)

Which of the jobs pictured above should/could be analyzed using the Critical Task Analysis technique?

(See Page 199)

Sugar refinery explosion

(See Page 223)

Before these job tasks were started, how would you have communicated the hazards to the people performing the work?   (See Page 244 and 10)

"Those who cannot remember the past are condemned to repeat it."

George Santayana

(See Page 253)

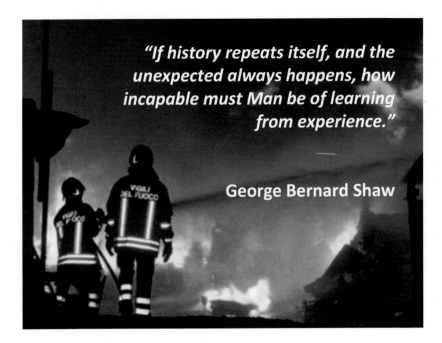

"If history repeats itself, and the unexpected always happens, how incapable must Man be of learning from experience."

George Bernard Shaw

(See Page268)

Incorrect or unsafe use of ladders may also occur in process plants. In some situations, portable ladders are used instead of platforms and fixed ladders to access equipment (Figure 4-33). If portable ladders are used, they must be industrial grade. As a result, the operator could be performing a task (e.g., valve operation) from the ladder that requires him to abandon a 3-point stance, creating the potential of falling. Plants should check all jobs where ladders are being used. Falls are the potential result of improper ladder usage. Improper ladder usage includes:

- Not tying off straight ladders
- Not extending straight or extension ladders at least 3 feet above a landing to step off the ladder
- Climbing or working from step ladders above the next-to-top step
- Improper straight ladder angle
- Not keeping your body between the ladder rails, etc.

Figure 4-33. Use of portable ladders to perform tasks can result in fall hazards

4.7   THERMAL HAZARDS

Thermal hazards can include both heat and cold hazards. This section discusses temperature hazards associated with the process or chemical properties of process materials.   Hazard associated with ambient temperature extremes are discussed in Section 4.17.

4.7.1   Heat

Fire and heat hazards include temperatures that can cause burns to the skin or damage to other organs.

Common injuries in a process facility are burns from contact with hot process lines and steam lines.   In some cases, the plant is not designed to allow operators safe access to or between (Figure 4-34) equipment by having the hot lines insulated or ensuring that the access clearances are substantial enough to reduce the potential of contact. Many companies have engineering standards that require burn protection insulation for hot lines and process equipment that are within reach of workers.

Tight clearances present other hazards. In the event of a sudden release, tight congestion can impede the safe egress of workers from the area.

Figure 4-34.   Access clearances are often not adequate enough to prevent operators from contacting hot lines

Inadequate access to equipment can cause operators to stand on lines to reach valves or to perform maintenance (Figure 4-35). This can expose operators to hot lines and can damage insulation (Figure 4-36).

Figure 4-35. Lack of access encourages operators to stand on lines for access to equipment

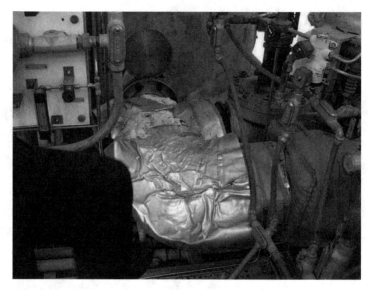

Figure 4-36. Continually using lines for access damages insulation

4.7.2   Cold

Exposure to extremely cold process equipment or materials can lead to embrittlement and subsequent equipment failure, as well as personnel injury. Cold hazards in process facilities include:

- Flashing liquids, like propane or ethylene
- Cryogenic processes

4.8   VIBRATION HAZARDS

Vibration hazards may lead to material fatigue and subsequent failures. Extreme vibration of lines, pipes, cables or supports needs immediate attention to avoid catastrophic ruptures, weld and flex-stress failures.

Excessive vibration combined with inadequate clearances can cause chafing of lines and cables. This may result, over time, in weakened line walls or cable insulation that can ultimately cause failure.

4.9   MECHANICAL FAILURE HAZARDS

Mechanical failure hazards typically occur when loads on devices exceed designed capacity or when equipment is inadequately maintained. Weight capacity for personnel platforms, mezzanines, ladders, devices and equipment should be prominently posted.

Good practice is to operate equipment within specified operating limits. When equipment is operated outside specified limits, it may fail. Figure 4-37, for example, shows the results of a mechanical failure hazard.

Figure 4-37.   Motor bearing failure causes couplings to fail, damaging the guard

It is possible for plant workers to exceed the lifting capacity of overhead cranes. The cranes may not be labeled with safe lifting capacities. The capacities at different lift distances (Figure 4-38) may not be listed. The loads may be unknown or difficult to calculate correctly. Whatever the reason, crane failures can sometimes cause catastrophic events that incur injury and major loss. For large loads, a formal lift study must be conducted by a qualified engineer.

Figure 4-38. Crane lifting capabilities vary with the horizontal distance at which the load is lifted

4.10   MECHANICAL HAZARDS

Mechanical hazards occur from machines and equipment that have powered mechanical moving parts that have the potential for crushing, cutting, tearing, or shearing if the human body comes into contact with these moving parts.

There are numerous ways in which workers can become exposed to a mechanical hazard that has the potential for injury. Lack of guarding is one of the most frequent ways in which plant workers are injured (Ref. 4-2). Proper guarding helps protect workers from exposure (Figure 4-39).

Figure 4-39. Proper equipment guarding can help prevent severe injuries

Pumps and other rotating equipment need to be checked for proper placement of rotating shaft guards. Rotating shafts or couplings present major potential for "grabbing" clothing and pulling arms or legs onto the shaft. Openings at the ends of the guard between a pump on one end and its motor on the other end of the guard need to be about ½" or less.

Another potential exposure occurs when workers must remove guards to service equipment with powered moving parts. A lockout-tagout program is essential to identify the sources of energy that can cause the machine or equipment to operate and lock out these sources of energy while the servicing is in progress. Locking out these energy sources and "bump testing" equipment to prove "zero energy" is necessary prior to servicing.

4.11   CORROSION HAZARDS

Corrosion in a plant may occur very slowly over time, and is often hidden from the casual observer by other process equipment, insulation, labeling, etc. This corrosion can often be where operators place their hands (Figure 4-40) or brush against as they navigate the plant. Corrosion may also exist inside process equipment where it is in direct contact with process fluids or under insulation; this corrosion cannot be easily detected without the use of specialized methods and equipment. Where internal corrosion is suspected, equipment should be thoroughly inspected at the next planned maintenance outage. Severe corrosion can lead to loss of containment resulting in a significant fire and explosion.

Figure 4-40. Corroding pipes can lead to loss of containment

4.12   NOISE HAZARDS

Noise levels (>85 decibels (dbA) 8 hour time-weighted-average (TWA)) that result in hearing damage or inability to communicate safety-critical information create a hazard. Areas within the plant site should be analyzed for the potential for hearing damage and, if they exceed requirements, mitigation should be implemented. The mitigation can be an engineering redesign of the equipment or surroundings to reduce exposure to workers. It could also be the use of personal protective equipment (PPE) to reduce the effects of noise at the ears. If PPE is considered, workers must be warned of the presence of excessive noise. The signage shown in Figure 4-41 is acceptable, but that shown in Figure 4-42 is not.

Figure 4-41. Acceptable hearing protection signage

Figure 4-42.   Unacceptable hearing protection signage

4.13   RADIATION HAZARDS

4.13.1 Ionizing Radiation Hazards

Ionizing radiation hazards that cause injury (tissue damage) by ionization of human cells include:

- Alpha
- Beta
- Gamma
- Neutron particles
- X-rays

Ionizing radiation is mutagenic (cell changing) to workers. Workers must be warned of the danger of ionizing radiation through internationally accepted signage (Figure 4-43).

Figure 4-43. Internationally accepted DANGER sign for ionizing radiation

4.13.2   Non-ionizing Radiation Hazards

Non-ionizing radiation is radiation that has enough energy to move around atoms in a molecule or cause them to vibrate, but not enough to remove electrons.   This radiation is not typically mutagenic (cell changing) to workers. However, this radiation can adversely affect instruments such as heart pace makers and can create rapid and significant heating of human tissue. Examples of this kind of radiation are:

- Ultraviolet
- Visible light
- Infrared
- Microwaves
- Lasers

Workers must be warned of the danger of non-ionizing radiation through internationally accepted signage (Figure 4-44).

Figure 4-44.   Internationally accepted DANGER sign for non-ionizing radiation

4.14   IMPACT HAZARDS

Impact hazards can be defined as a mass that strikes the body causing injury or death. (Examples are falling objects, projectiles and obstructions that are hit by workers). Workers in the normal course of their jobs may walk into obstructions or walk into the path of a solid moving object. In any situation involving the transfer of momentum, a serious injury may result.

Preventing items from falling from heights is the most effective mitigation against impact hazards. For the rare times that items do fall, workers can sometimes be protected by the proper use of personal protective equipment. It is important that policies are in place to ensure workers are wearing PPE that is specified for the hazards they are exposed to.

In addition to being struck by falling objects, workers are also exposed to obstructions that they can strike their bodies against during the normal course of their work activities (Figure 4-45 and Figure 4-46). In many cases, the obstructions are in the normally traveled paths of plant workers (rounds paths). In these cases, the obstruction must be highlighted and well lit 24 hours per day. If at all possible, the obstruction should be removed or another route designated for travel.

Figure 4-45.   Head obstruction under process vessel

Figure 4-46.   Body obstruction in the designated path
through the plant

4.15   STRUCK AGAINST HAZARDS

Struck against hazards include injury to a body part as a result of coming into contact with a surface in which action was initiated by the person (e.g., a screwdriver slips).

Tools slip and hurt people. Often it is because they are not being used for the purpose that was intended (Figure 4-47), or they are not designed properly. In any case, when tools slip or break, there is a reasonable likelihood that workers will get hurt.

Figure 4-47. Tools not being used for their designed purpose
(See color insert)

4.16   VISIBILITY HAZARDS

Visibility hazards include lack of lighting or obstructed vision that results in an error or other hazard.

Inadequate or inoperative lighting makes it difficult to enter an area and perform even routine tasks.   Inadequate lighting may also create hazards for workers by obstructing their view causing them to take incorrect actions.

In addition to lighting, the plant environment can affect visibility. Steam from heated equipment or fog from chilled equipment; for example, can obscure equipment (Figure 4-48) creating dangerous working environments.

Figure 4-48.   Steam obscures process equipment creating a hazardous working environment

4.17   WEATHER PHENOMENA HAZARDS

Weather hazards can affect workers who must navigate the plant site and work on equipment installed outside. Weather hazards can result in workers injury. Weather can also have a serious effect on process equipment especially moving equipment or relief device discharge pipes.

4.17.1 Temperature Extreme Hazards

Temperature extreme hazards include temperatures that result in heat stress, exhaustion, or metabolic slow down such as hypothermia.

In some locations, both high and low temperature extremes cause body stress to plant workers. In the warmer climates, high temperatures restrict work levels and work times to ensure that core temperatures do not rise to a point where operators will suffer from heat stress.

Plant policies should allow plant workers to work safely in all outdoor environments and should provide the clothing and equipment (e.g., cold vests for heat, arctic clothing for cold) that will keep workers safe from injury and illness.

4.17.1.1 Freezing Temperatures

In cold climates, the issue is working outside in very cold temperatures. In these conditions, the hazards are:

- Cold stress and hypothermia because of the lack of proper clothing and gloves (Figure 4-49)
- Increased probability of musculoskeletal injuries in cold weather
- Falling ice from process structures and buildings
- Equipment failure (including critical safety equipment)
- Ice-coated slippery surfaces

Extremely cold temperatures can adversely affect equipment performance and may lead to premature failure. Lubricants may freeze and bearing life may be reduced. Water and other process fluids may freeze in dead zones, causing expansion and physical damage. Finally, measures taken to protect against extreme weather exposures may introduce additional hazards.

**Figure 4-49.   Cold weather hazards increase the potential
for illness and injury**

Extremely low temperatures present exposure hazards to workers from frostbite and to the skin and stress on the respiration system (Figure 4-50).   When this weather is predicted, protection is provided through well designed and layered clothing as well as masks to provide warmth to air as it is inhaled.   Workers should work in pairs with each person observing and checking with the other as to their well being.   Breaks should be taken periodically to warm up and check fingers, toes and ear lobes for signs of frostbite. Wet clothing should be changed promptly.

Figure 4-50. Very cold temperatures can affect working performance

4.17.1.2 High Temperatures

Extremely high temperature can result in heat exhaustion or more serious outcomes. Workers should drink plenty of water, wear proper clothing and head protection against the sun, and take periodic breaks to cool off. Workers should also observe each other for signs of confusion which can indicate the onset of extreme heat stress (Figure 4-51).

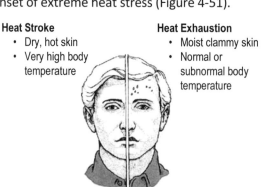

Heat Stroke
- Dry, hot skin
- Very high body temperature

Heat Exhaustion
- Moist clammy skin
- Normal or subnormal body temperature

Figure 4-51. Signs of heat stress

4.17.2 Hurricane

Hurricanes are typically forecast by the weather services so that planned preparation can be made to secure operations or shut down and evacuate as the predicted storm level indicates. Many of the United States refineries and chemical plants are located around the Gulf of Mexico, a primary location for hurricanes. Offshore facilities also have to prepare for shutdown and evacuation in the event of hurricane or other severe weather conditions (Figure 4-52).

Figure 4-52. Offshore platform in hurricane conditions *(See color insert)*

4.17.3 Flood

Floods are also typically forecasted by the weather services. If these exposures are a potential, planned preparations should be undertaken to safely secure operations prior to the rise of the flood waters (Figure 4-53).

Figure 4-53.   Process facilities impacted by flooding *(See color insert)*

4.17.4   Wind

High winds are associated with hurricanes and tornadoes.   Tornado-producing weather conditions are typically forecast by the weather services; however, unlike hurricanes or floods, the predictions are less exact and the timing to take preparatory actions is much less. If tornados are an expected weather event, planning is required to quickly secure operations as best as possible and seek shelter in appropriate strength locations.   Winds can impact and damage process facilities (Figure 4-54). Insulation and cladding on process equipment is particularly vulnerable to high wind conditions and may become airborne where it represents a physical hazard to people and equipment. In addition, wind gusts and high winds can upset crane loads or make it too dangerous to climb towers.

Figure 4-54.   Wind damage at polymer facility *(See color insert)*

4.18   REFERENCES

4-1.   U.S. Chemical Safety and Hazard Investigation Board. "Static Spark Ignites Flammable Liquid during Portable Tank Filling Operation". Case Study. September 2008. www.csb.gov

4-2.   Attwood, D.A. "Understanding and Dealing with Human Error in the Oil, Gas and Chemical Industries." Proceedings, CCPS Meeting, San Antonio, TX. 1998.

5

EVALUATE HAZARDS

This chapter focuses on methods for identifying and evaluating hazards in the workplace (Figure 5-1). The methods discussed are a compilation of techniques that are routinely used by process industries to help identify hazards in a process, a chemical, or a particular job task. When selecting a hazard evaluation method or technique, it's important to remember that these are tools and "one-size does not fit all". Customization of the evaluation method or technique may be required.

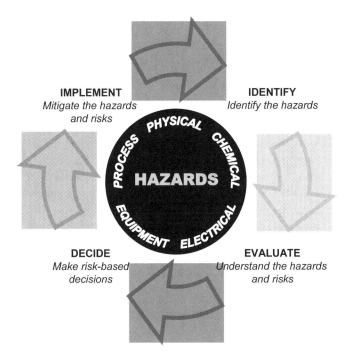

Figure 5-1. Hazard management process

Identifying hazards often depends on our ability to recognize the hazard. While it might seem sufficient to send workers out into the facility to identify hazards in their work place, it is beneficial to provide them with a systematic method for hazard identification.

INCIDENT – PIPELINE EXPLOSION

A buried gas pipeline was exposed adjacent to a construction site. One morning, the workers complained of a foul odor and immediately evacuated the area. This was a wise decision. A group of emergency responders soon arrived and began to survey the situation; a gas leak was readily apparent to these workers. Because the gas line was part of a major utility supply network it was difficult to depressurize and isolate the line. Within one half hour of the reported leak, the line ruptured resulting in an intense explosion and fireball.

Twenty-three workers (mostly responders) were killed and over 100 others were injured. Property damage was extensive over a wide area. The failure of the line was the result of brittle fracture; gas expanding and escaping had cooled the line to a very cold temperature level. An investigation revealed that the line had been damaged one week earlier by construction equipment but the incident had not been reported. All mechanical defects must be reported and dealt with to ensure that they do not contribute to major incidents. Emergency responders must follow the same rules as other workers and must be aware of hazards in the workplace.

What could have been done to prevent this incident?

Who had the responsibility to take action after the initial damage incident?

This chapter provides a description for a number of hazard identification methods that have been developed in the process industry. They can be categorized into three groups:

1. Field Surveys: Methods that are designed to collect data in the field either through walk-through surveys using custom-designed checklists or through the observation of workers as they perform operations or maintenance tasks.
2. Pre-job Assessments: Methods that are designed to evaluate the job site to identify hazards in the area of interest.
3. Facility Assessments: Methods that are designed to identify hazards either before the facility is designed or in an existing facility with the thought of improving the design of facilities or equipment.

These methods can be used to identify hazards associated with facilities and job tasks, like those pictured below. *(See color insert)*

The goal of this chapter is to provide the reader with:

- A full description of each of the methods
- Literature references for published methods
- An overview of when and how each of the methods can be used

The reader should appreciate that each generic method is typically customized by a user to fit the characteristics of a particular facility. While the description provided below is typical, the actual method, when used in the field, may vary from the way it is depicted here.

5.1   FIELD SURVEYS

Field surveys and worker observations are proactive methods for identifying hazards and preventing incidents. This section reviews field survey methods that are used by process industries.

5.1.1   Behavior Observation

Behavior is defined as what workers do or say - i.e., their actions - not what they think or feel. Thus, behavior is an objective observable concept. This section reviews the use of behavior observation as a method of identifying workplace hazards.

Observing the behavior of operators and maintenance workers can provide insight on:

- Errors that are committed during the performance of the task
- Time required to perform each activity
- Difficulty or ease with which the task is performed
- Improved ways of performing the task or alternative methods that could be used to perform more safely or efficiently
- Quality of procedures

Observations can be performed in three different ways (Ref. 5-1):

- Direct Observation: where an observer deliberately observes another person (or persons) performing a task (0). This is arguably the most frequent method used and is key to behavior-based safety programs.

> **What should you do if you observe one of your coworkers continuing to violate procedures and take shortcuts?**

Figure 5-2.   Observing behavior of a worker performing a task

- Indirect Observation: where those being observed do not know they are being observed. Traffic and pedestrian surveys fall into this category.
- Participatory Observers: where the observer is the one performing the task.

For each type of observation scheme, preparation is the key. Some items to consider during preparation include:

- *What activities will be observed?* Generally, people are observed doing work. Often the types of work observed are critical tasks - those that have caused injuries, loss, or environmental releases in the past or have the potential for doing so in the future.
- *What is the instrument that is used for observation?* Critical behaviors are often identified from near-miss data or prior observations. Behavior-based safety practitioners (Ref. 5-2), (Ref. 5-3), (Ref. 5-4) and (Ref. 5-5) recommend using checklists of the critical behaviors of interest. Often, critical behaviors are identified by workers in brainstorming sessions. Then observations are conducted of the operations or maintenance tasks that contain critical behaviors.

As noted by (Ref. 5-5), "By itself, one observation may not be a good measure of safety performance. During one observation, behaviors will be missed." In other words, several observations may be necessary in order to capture the at-risk behavior. Each observation checklist should describe what the observers should be looking for. Clearly, there could be a disconnect between the visual scene and its description on paper that could cause observers to miss or misclassify the observation. Checklists should be designed to be as visual as possible. For example when observing a valve operating task, the hazard could be an unstable standing surface as shown in Table 5-1.

Table 5-1.   Illustrating the hazard under observation

Checklist Item	Visual Representation	Safe Behavior (√)	At-risk Behavior (√)	Comment
Worker stands on stable surface (grade or platform) when operating valve.	Example of an unstable standing surface.		Using piping or equipment to stand on to access a valve is at-risk behavior and can leave the operator with injuries if they slip and fall.	If the valve is frequently operated, a permanent solution must be implemented to allow the operator access to the valve. Solutions could include a permanent step or the valve could be equipped with a chain-operator.

- *Who is being observed?* A random sample of workers that cuts across all shifts and all times of the day should be observed.   It is essential that those being observed are representative of the population under consideration. Students or temporary workers, for example, performing a skilled task as study subjects do not behave the same as skilled operators performing the same task.
- *Who is making the observations?* Typically, other workers, supervisors or outside consultants make observations. It is essential that the observers are trained in observation techniques and know what they are looking for. It is also imperative that those being directly observed know why and how the data will be used. They should also feel comfortable with the person observing them so that they perform the job task the way they would if they were not being observed. The observation activity may affect the data collected because people may change their behavior when they know they are being observed.

- *What data are being collected?* Decide early what variables will be collected and prepare for them. For example, potential errors need to be specified so the observer can be trained to 'see' them and they can be coded on a form. In addition to errors, determine how long it takes to perform the task or how many people are required to perform it.

5.1.2   Facility Walkthrough Checklists

Facility walkthroughs are a common method of looking for hazards and are typically conducted by operations and supervisor/management:

- Operations facility rounds - Facility rounds are conducted by operators several times per shift to monitor equipment operation and observe that the equipment is operating satisfactorily, e.g., no leaks, no vibration. The rounds may be guided by a data collection sheet on which values of process variables are recorded.
- Facility management walkthroughs - Facility walkthroughs can also be conducted by facility management to look for safety hazards such as broken insulation, steam leaks, tripping hazards, water on grade, hot water dripping on walkways, poor housekeeping, broken equipment, missing plugs, broken lighting, etc.

Often walkthroughs are informal and issues are written as notes to be left with the shift supervisor or added into shift logs. Informal observations that are gathered during facility walkthroughs are encouraged. However, walkthroughs are most efficient when guided by a checklist of some type (Figure 5-3). Checklists are defined (Ref. 5-6) as "... a carefully compiled, comprehensive list of protective measures, material properties, hazards or 'good practice' design features that have been compiled by experienced workers for a particular application."

Figure 5-3.   Completing a checklist during a plant walkthrough

PLIBEL is an ergonomic hazard checklist that was designed to highlight musculoskeletal risks in connection with workplace investigations (Ref. 5-7). The checklist was designed so that items ordinarily checked in a workplace assessment would be listed and linked to one of five body regions. When an ergonomic hazard is observed, the numbered area on the form is checked or a short note is made.

Table 5-2 illustrates an ergonomic checklist that was developed by the U.S. Department of the Navy (Ref. 5-8). The table provides a backup illustration for each question to help the checklist user interpret the question and decide whether the hazardous situation exists or not. Table 5-2 illustrates the use of recommended safeguards to mitigate the hazard if it is observed.

Table 5-2.   Example ergonomic checklist

Posture and Layout	Yes	No	Possible Solutions to Reduce the Risk of Injury
Are objects handled between knuckle and shoulder height?			— Minimize the number of times the load is lifted below the knees or above the shoulders — Provide a mechanical aid, like a hoist or cart — Frequently used objects should be within easy reach, close to waist height if possible
Are objects within arm's length, allowing the worker to reach them without bending their back?			— Allow workers to use different postures and muscle groups through job rotation — Provide a mechanical aid, like a hoist or cart — Frequently used objects should be within easy reach
Is the task performed in an open space, allowing the worker to freely move feet and arms?			— Allow workers to use different postures and muscle groups through job rotation — Adjust height of the workstation to optimal working height — Provide a mechanical aid, like a hoist or cart

The examples in Table 5-3 provide several approaches for designing a facility walkthrough hazard identification checklist. In summary, the checklist should:

- Categorize the item into a hazard group, such as maintenance, operability or design so that it can be addressed by the appropriate department
- Illustrate the hazard to help the user make a decision
- Provide suggestions to mitigate the hazard

Table 5-3 presents a process hazard checklist that could be used during a facility walkthrough that is based on the above recommendations. This is not all inclusive. This example can be used to build your own site-specific checklist.

Table 5-3.   Walkthrough checklist

Checklist Question	Illustration	Present (√)	Hazard Category			Potential Mitigation
			Maint.	Opr.	Design	
1. Potential adverse outcome from inadvertent leaning against control panel, switches, etc.		√		√	√	Plastic covers over switches. Admin rule no sitting on control board.
2. Labels on critical switches, valves, piping and vessels inadequate		√			√	Replace label.
3. Valve access inadequate		√	√	√		Provide step/ platform.

Table 5-3. Walkthrough checklist (continued)

Checklist Question	Illustration	Present (√)	Hazard Category			Potential Mitigation
4. Electrical connections and enclosures loose or open		√	√			Replace junction box with large box and install cover.
5. Insulation inadequate		√	√	√	√	Provide insulation to protect from freezing.
6. Sample point inaccessible		√	√	√		Provide piping to allow access.
7. Eye-wash station is obstructed and difficult to access		√	√	√		Move equipment and paint surface to indicate area left open.
8. Corrosion leading to failure		√			√	Clean, inspect, repair.

WARNING – VACUUM TRUCKS

Vacuum trucks are commonly used in heavy industry to collect and dispose of fluid and solid waste. Sometimes the physical and chemical properties of the waste material are not disclosed to the operators of the equipment.

One winter, two workers were proceeding down a roadway in a vacuum truck loaded with hydrocarbon sludge.

A passing motorist signaled the driver that material was leaking from the drain port at the rear of the truck. The workers stopped the vehicle and attempted to tighten the drain valve. However, it was frozen in the partly opened position. One worker lit a gas torch to apply heat to the frozen valve. The tank immediately exploded ejecting the rear bulk head and killing both workers. The material in the tank contained flammable hydrocarbon that ignited from the open flame. The flame flashed back into the tank causing the contents to explode.

Vacuum truck incidents are all too common in industry. It is important to know the contents of a vacuum truck and apply strict safety measures. This is especially important when different material have been transported.

Workers should know the hazards of the materials.

What other hazards contributed to this incident?

5.2   PRE-JOB ASSESSMENTS

5.2.1   Job Hazard Analysis

A Job Hazard Analysis (JHA) is a technique that is used to identify and assess the hazards in a job before they occur. Other terms used to describe this technique are Job Task Analysis (JTA), Job Safety Analysis (JSA) and Job Hazard Breakdown (JHB). The JHA focuses on a specific job and examines the:

- Steps required to perform the job
- Relationship between the job and the worker(s), methods used by the worker(s) and the work environment
- Risk of injury involved in the task and specific actions to reduce the risk

This section demonstrates how to perform a JHA. The information contained herein can be found in more detail in three major references: (Ref. 5-9), (Ref. 5-10) and (Ref. 5-11). In its simplest form, the JHA process consists of four steps:

Step 1: Identify hazardous jobs
Step 2: Determine the risk of injury for each job selected
Step 3: Prioritize jobs for analysis based on their risk of injury
Step 4: Analyze the job to determine the detailed hazards, identify the best actions to eliminate or reduce the hazards, and plan to apply the actions in the field

JHAs should also be conducted on:
- Newly Established Jobs: Due to the lack of experience with new jobs, hazards may not be understood.
- Modified Jobs: As part of your management of change program, each modified job should be reviewed for new and existing hazards.

Which type of field survey would you use to identify the hazards pictured here?

5.2.1.1 Identify Hazardous Jobs

The purpose of the JHA is to analyze hazardous jobs, not to identify facility hazards. Hazardous jobs can be identified in several ways:

- From the results of incident investigations
    - Jobs where incidents occur frequently
    - Jobs with the potential for severe injuries
- By workers at job review meetings based on their work experience

The key to identifying hazardous jobs is to first understand what process hazards are. Incident investigations and job review meetings are ineffective, unless the people involved in them understand what the hazards are and can identify the jobs that contain them. In reality, almost every job contains at least one of the hazards in Table 5-4. The first pass at identifying hazardous jobs is to select those that are considered to be at higher risk.

Table 5-4.   Common process hazards

Hazards	Hazard Descriptions
Asphyxiates	Asphyxiating atmospheres such as excessive nitrogen, carbon dioxide, or carbon monoxide or drowning or smoke
Biological hazards	Conditions leading to exposure to biological hazards such as the flu, HIV or contaminated food or water
Chemical (corrosive)	A chemical that, when it comes into contact with skin, metal, or other materials, damages the materials. Acids and bases are examples of corrosives
Chemical (explosions)	A chemical that when exposed to heat or pressure results in explosion
Chemical (flammable)	A chemical that, when exposed to a heat ignition source, results in combustion
Chemical (toxic)	A chemical that exposes a person by absorption through the skin, inhalation, or through the bloodstream that causes illness, disease, or death
Co-located hazards	Conditions such as pressure relief valves and process stacks
Dynamic Situation Hazards (struck by (mass acceleration)/struck against)	Equipment with moving or rotating parts, accelerated mass that strikes the body causing injury or death (examples are falling projectiles)
Electrical (arc/flash)	Use of electrical power that results in electrical overheating or arcing to the point of combustion or ignition of flammables, or electrical component damage
Electrical (shock/short circuit)	Contact with exposed conductors or a device that is incorrectly or inadvertently grounded, such as when a metal ladder comes into contact with power lines Lightning discharge (spark) to the ground that results in the ignition of flammables or damage to electronics or the body's nervous system
Entrapment	Blockage or access/egress to escape routes or the snagging of lines/umbilicals or soil collapse in a trench
Environmental hazards (weather phenomena (snow/rain/wind/ice), sea state or tectonic)	Natural environmental conditions such as earthquakes, wind storms or violent seas

Table 5-4. Common process hazards (continued)

Hazards	Hazard Descriptions
Ergonomics	A system design, procedure, or equipment that is error-provocative. (A switch goes up to turn something off) Damage of tissue due to over exertion (strains and sprains) or repetitive motion Vibration that can cause damage to nerve endings Skin, muscle, or body part exposed to crushing, caught-between, cutting, tearing, shearing items or equipment Temperatures that result in heat stress, exhaustion, or metabolic slow down such as hypothermia Lack of lighting or obstructed vision that results in an error or other hazard Temperatures that can cause burns to the skin or damage to other organs
Explosives	Material such as explosives, detonators or perforating gun charges that are designed to explode
Hazards associated with differences in height	Conditions that result from differences in height such as falls from scaffolding, ladders, platforms towers stacks, or objects falling while being handled onto people, equipment or process systems
Heat/open flame	Equipment with open flames such as re-boilers, fired heaters or flares.
Material (explosive)	Material that under certain conditions will explode, e.g., dusts
Material (flammable)	Conditions leading to the spontaneous ignition of materials such as wood planks, paper or rubbish or spontaneous ignition of pyrophoric materials such as metal scale from vessels in sour service
Mechanical/vibration (chaffing/fatigue)	Material fatigue, chaffing, vibration, pneumatic
Noise	Conditions under which noise levels can result in hearing damage or inability to communicate safety-critical information
Objects under induced stress	Objects under tension or compression such as guy and support cables or spring-loaded devices such as relief valves
Pressure hazards	Gases under pressure that may result in an uncontrolled high pressure release
Psychological hazards	Conditions that lead to fatigue or stress

Table 5-4.   Common process hazards (continued)

Hazards	Hazard Descriptions
Radiation (ionizing)	Ionization of cellular components such as alpha, beta, gamma, neutral particles, and X-rays
Radiation (non-ionizing)	Ultraviolet, visible light, infrared, and microwaves that cause injury to tissue by thermal or photochemical means
Security-related hazards	Conditions such as assault and theft or equipment exposed to sabotage, pilferage and terrorism
Temperature extreme	Hot or cold conditions, material or equipment

One method for doing this in a meeting format is to use a "nominal group" technique (NGT) that could be conducted as follows:

- Workers are introduced to the list of potential hazards from Table 5-4 (adapted from OSHA 3071, 2002 and ISO 17776, 2000), examples of hazards in the facility are identified and jobs that contain hazards are discussed. At this point, a revised list of hazards is prepared.
- Workers are then asked to identify jobs that contain hazards on the revised list. This can be a brainstorming session, or they can be written on 'sticky notes' and pasted to a board (Figure 5-1). Only one hazard from the list is dealt with at a time.

Figure 5-1.   Using sticky notes to brainstorm hazardous jobs

- The hazardous jobs are reviewed and duplicates are eliminated.
- Each participant then ranks each job in terms of its risk for injury. For example, if ten jobs are listed, the highest risk job would be given a ranking of 10 and the lowest a ranking of 1. This is illustrated in Table 5-5.
- Rankings are added across participants for each job, and the jobs are re-ordered based on the total ranks.

The result is a subjective ranking of hazardous jobs. Only those considered most hazardous will enter subsequent stages of the JHA.

Table 5-5.   Hypothetical ranking sheet from an NGT session

Job	Scores from 5 Participants (10 = High Risk, 1 = Low Risk)	Total of 5 Scores	Average Risk Score (5 participants)	Rank (10 = Highest Risk)
Conduct In-Line Inspection (ILI) of 4-inch lines - steam generator	8, 10, 7, 7, 9	41	8.2	8
Remove cartridge filter - amine service	9, 9, 6, 10.8	42	8.4	9
Maintain emulsion pump	5, 6, 1, 8, 6	26	5.2	6
Clean fire water knockout drum/treater	4, 3, 5, 6, 2	20	4	3
Clear ice plug from line	3, 5, 4, 5, 7	24	4.8	5
Sample amine solution and add make-up water as needed	10, 8, 9, 9, 10	46	9.2	10
Blow down waste heat re-claimer boiler	1, 4, 2, 4, 4	15	3	2
Re-light incinerator	2, 1, 3, 1, 1	8	1.6	1
Re-start pre-lube engine and compressor	6, 7, 10, 3, 5	31	6.2	7
Respond to seal leaks on treater pumps	7, 2, 8, 2, 3	22	4.4	4

5.2.1.2   Analysis and Job Prioritization

The analysis is typically performed by a group of operators and maintenance technicians who are familiar with each job being considered.

The risk of injury for each high-ranking job from the above list is categorized in terms of likelihood of occurrence and severity of injury. The process is similar to the one described in detail in Section 5.3.2 Critical Task Identification and Analysis. The result is a prioritized list of tasks for analysis. Typically, the highest risk jobs are analyzed first.

5.2.1.3 Job Hazard Analysis and Follow-up

The analysis portion of the Job Hazard Analysis (JHA) consists of three discrete phases that can be applied to new or existing process or job tasks:

- List the job steps
- Identify the hazards involved in each step
- Identify safeguards that mitigate the hazard

The process typically uses a JHA form like the one reproduced in Table 5-6. As in other analyses, the JHA is conducted by a group of operations and maintenance technicians who are intimately familiar with the jobs. They will be able to accurately break the jobs down into their constituent parts and they can identify, through experience, what the hazards are. The group could also include an engineer, a safety specialist, a process safety specialist and a hygienist who can help identify practical safeguards to mitigate the risk.

Table 5-6.   Job Hazard Analysis worksheet

Job: Conducting In-Line Inspection (ILI) of 4-inch Lines		Date: 03-19-09		Attendees: Operations Team Members		
Job Steps		**Hazards**		**Safeguards**		**Priority (H/M/L)**
1. Shutdown and depressurize and cool generator		1.1	Burns from steam isolation valves	1.1.1	Ensure steam isolation valves are pointed away	M
		1.2	Physical stress injuries from reaching	1.2.1	Ensure valves are at grade level or reachable from a platform	M
2. Drain lines		2.1	Exposure to line contents	2.1.1	Connections to drain hose	L
				2.2.1	Valves are reachable	L
		2.2	Stress from opening valves			

Table 5-6.   Job Hazard Analysis (continued)

Job: Conducting In-Line Inspection (ILI) of 4-inch Lines	Date: 03-19-09		Attendees: Operations Team Members	

	Job Steps		Hazards		Safeguards	Priority (H/M/L)
3.	LOTO - blinding generator, fuel gas, combustion air, power, steam header	3.1 3.2 3.3 3.4	Heat exposure Access to equipment Equipment removal Lifting	3.1.1	Allow time for process to cool	H
				3.2.1	Ensure break-line locations are accessible	H
				3.3.1	Use spec blinds	H
				3.4.1	Provide utility air for air wrenches	H
				3.5.1	Provide lifting lugs and lifting equipment for all items weighing more than 102lbs	H
4.	Remove spool pieces	4.1 4.2 4.3	Lifting Access to spooling area Lighting	4.1.1	Provide monorail for lifting devices and lifting lugs	M
				4.2.1	Provide cement flooring for drains	M
				4.3.1	Provide adequate lighting	M
5.	Hook-up sender/ receiver	5.1 5.2 5.3	Lifting Access Lighting	5.1.1	Install permanent sender/receiver	L
				5.2.1	Eliminate need for scaffolding	L
				5.3.1	Provide adequate lighting	L
6.	Circulate water from equipment to generator	6.1 6.2	Select wrong ILI device Hammer door operation	6.1.1	Code ILI device for proper selection	M
				6.2.1	Provide tools specific to hammer door operation	M
7.	Launch ILI device	7.1 7.2	Valves operated out of sequence Leaking door seal	7.1.1	Follow correct pressure build-up procedure	H
				7.1.2	Code valves to operate in proper order	H
8.	Monitor pressure on pump	8.1	Valves operated in correct sequence	8.1.1	Ensure that PSV cannot be set above design pressure	L
				8.1.2	Code valves for proper operation	L

The key at this stage is to accurately break the jobs into their component steps. The steps should:

- Begin with an action verb, e.g., shutdown generator, drain lines, etc. (Avoid the passive tense, such as "The lines should be drained.")
- Be described in enough detail to permit the team to identify individual hazards. If the steps are too detailed, the team can become bogged down in trivial issues, and if they are not detailed enough, hazards can be missed.
- Be recorded before continuing to the hazard identification (Column 2, Table 5-6).
- Be numbered. The numbering system should be consistent across columns as noted below.

Once each of the steps has been identified and verified by the team, the analysis can continue. For Step 1, the existing and potential hazards associated with the activity should be identified (Column 2, Table 5-6). A copy of the modified hazard table should be distributed to the participants for them to review at each stage. Each existing or potential hazard should be numbered consecutively. The first number is that of the step, e.g., for Step 1, the hazard numbers would be 1.1, 1.2, 1.3, etc.

After identifying each existing or potential hazard associated with a step, the potential safeguards that could be applied to each hazard to reduce the risk should be listed (Column 3, Table 5-6). This is a brainstorming exercise. No safeguard is unacceptable. Each safeguard is numbered consecutively, e.g., for hazard 1.1, the safeguards are 1.1.1, 1.1.2, etc.

A priority should be assigned to each intervention - High, Medium, or Low. The priority assignments are generally based on benefit to the task and cost of implementation. They could be determined simply on the consensus of the team, or they can be formally evaluated by assigning cost/benefit values to them (Ref. 5-12). Clearly, the safeguards that are most cost-beneficial in terms of risk reduction would be implemented first.

The following items (in decreasing order of robustness and reliability) provide measures for hazard control:

- Inherently Safer Designs: reduces or eliminates the hazards associated with materials and operations used in the process and this reduction or elimination is permanent and inseparable.
- Engineering Controls: A specific hardware or software system designed to maintain a process within safe operating limits, to safely shut it down in the event of a process upset, or to reduce human exposure to the effects of an upset.
- Management (Administrative) Controls: Strategies to eliminate or reduce exposure. This is done primarily by changing work practices, procedures, schedules (e.g., rotating jobs among workers).
- Personal Protective Equipment: Using PPE to establish a barrier between the hazard and the operator.

Inherently safer design, followed by engineering controls, is preferred over the others. But, redesigning or replacement can take time and effort. In the interim, one of the other control measures could be used until the preferred control can be implemented. Moreover, temporary measures might be implemented until a permanent fix is available.

5.2.2 Pre-Job Planning and Permitting

In many facilities, hazard identification, recognition and analysis includes many different programs that are conducted to meet varying needs. The JHA that is described above, for example, is conducted to identify hazards associated with process tasks. The JHA is often used to review jobs before they are performed to familiarize facility workers with the generic hazards. This is often part of a pre-job planning session.

Pre-job planning can be performed for every task that is performed in the facility. Often, however, it is conducted for those jobs that:

- Are non-routine
- Require a work-permit

If the job already has a JHA/JSA written for it, pre-job planning merely involves the work team and supervisor reviewing the JHA/JSA to ensure that the safeguards noted in the analysis are implemented and tailored to the specific job conditions. The hazards of changing a light bulb are very different in an office versus over an open vat of chemical, even though the job - changing the light bulb - is the same. Safeguards could include:

- PPE in addition to routine equipment, e.g., fall protection
- Hot or cold work permits
- Special equipment, such as cranes or barricades
- Special precautions, e.g., fire truck onsite

If the job does not have a JHA/JSA, then one must be prepared by the work team to ensure that hazards are identified and safeguards are adequate. Additional safeguards may be required to mitigate the hazard.

5.2.3   Ad Hoc Risk Assessment

Once workers arrive at the job site, they will often find that the work conditions that were assumed when the JHA/JSA was developed have changed or they encounter unexpected conditions. The site could be rain or snow covered, construction could be ongoing in the area, and minor changes could have been made to the equipment (Figure 5-4). Consequently, the JHA/JSA might not identify all the hazards on the job site.

Slippery floor from fluids

Step higher than normal

Ice and snow increase injury risk

Valve buried in mud

Figure 5-4.   Site conditions could change from those assumed in the JHA

The "Ad Hoc" Risk Assessment is carried out by the individuals who are performing the onsite work. The assessment is conducted at the job site just before performing work. The objectives of the process are to:

- Improve operators' hazard recognition and awareness abilities
- Retain awareness of the local hazards throughout the task
- Reduce the number and severity of incidents and illnesses in onsite activities
- Over the long term, to change operators' mindset such that the Ad Hoc Risk Assessments are done automatically without the need for a paper-based system

Many companies have developed a formal method for performing worksite risk assessments. In order to encourage site workers to conduct the analysis, companies have tried to keep the system simple and memorable. TAKE TWO... *for Safety*™ program, for example, was developed by DuPont and also marketed to other companies. The program steps are based on the mnemonic "TAKE" as illustrated in Figure 5-5.

T = <u>Talk</u> to all concerned about what is going to be done

A = Determine how my <u>Actions</u> can affect my safety and that of others

K = <u>Knowledge</u> of procedures and hazards

E = Determine whether I have the proper <u>Equipment</u> and whether it is in good working condition

Figure 5-5.   Illustrating the steps in the DuPont "TAKE 2" Program

Several process companies have developed their own programs complete with mnemonics to accomplish the same goal - ensure that the operators are aware of and remember the local hazards associated with their immediate task.   One such program developed by ExxonMobil Chemical is called SCAN© (Figure 5-6, modified from ExxonMobil).

FRONT

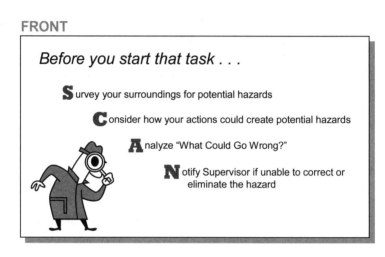

Before you start that task . . .

S urvey your surroundings for potential hazards

C onsider how your actions could create potential hazards

A nalyze "What Could Go Wrong?"

N otify Supervisor if unable to correct or eliminate the hazard

BACK

**Hazardous Conditions and Actions
You Can Ban**

When You First Remember to . . .

S. C. A. N.

Figure 5-6.   A last minute hazard risk assessment program

To use the SCAN© program, each worker is given a laminated card that provides the mnemonic and memory jogging slogans for conducting a last minute risk assessment.   Workers are trained on the proper use of the card.

The Ad Hoc Risk Assessment is not a substitute for a proper Job Hazard Analysis (JHA) program. In fact, it improves the program by ensuring that the unique onsite risks and hazards are identified, resolved and re-visited along with the common hazards that are outlined by the JHA and reviewed in the job planning sessions.

The Ad Hoc Risk Assessment works best if:

- Operators (including contractors):
  - Take responsibility for their safety and the safety of those they are working with
  - Consistently follow the Ad Hoc Risk Assessment process for onsite permitted jobs (especially for non-routine, high risk operations)
- Supervisors support the operators by:
  - Training operators and contractors in the technique
  - Encouraging operators to complete Ad Hoc worksite risk assessments to convince them that the technique is beneficial in reducing accidents and to promote a "want to do" approach rather than a "have to do"
  - Encouraging the completion of Ad Hoc worksite risk assessment to a consistently high standard
  - Committing to lead by example
  - Ensuring compliance by review and follow-up
- Facility management supports the program by:
  - Providing funding to enable the Ad Hoc Risk Assessment to be implemented
  - Stewarding the progress of the program
  - Ensuring that compliance is consistent across shifts

Here are some suggestions to ensure that Ad Hoc Risk Assessments are effectively conducted at the work site:

- Ensure that work site hazards are identified
- Provide enough time for the workers to complete the assessments
- Ensure that the workers understand the hazards

Consider, for example, the data presented in Table 5-7. These are a list of minor injuries (first aids and medical aids) that occurred over a one-year period at a North American chemical plant. Each person who suffered one of these injuries was an experienced plant operator and each injury was preventable.

How could an experienced person slip on pellets or strike his face while pulling with pliers?

Could an Ad Hoc Risk Assessment be conducted before the task was initiated have prevented these injuries?

Table 5-7.   Minor injuries as a result of lack of attention

Incident Type	Incident Classification		
	Hazard Identification	Equipment	Procedures
Slipped on pellets	X		
Cylinders fell on worker			X
Bicycle slipped on gravel	X		
Platform fell due to loose clips		X	
Bolt broke while tightening			X
Cardboard in left eye	X		
Caught right hand between two pieces of pipe	X		
Hit by high pressure water		X	

Table 5-7.   Minor injuries as a result of lack of attention (continued)

Incident Type	Incident Classification		
	Hazard Identification	Equipment	Procedures
Strain from door weight		X	
Loose step		X	
Cut hand while taping	X		
Laceration while installing grating	X		
Strike face while pulling with pliers		X	

The keys to injury prevention in each instance are:

- Was the hazard identified?
- What effect does the hazard have on you and others?
- How can you control the hazard? (What are the safeguards and barriers?)
- How do you remember the hazard while you are working?

Recalling the information on attention and memory from Chapter 3, what could be done to improve the Ad Hoc Risk Assessment and potentially prevent the injuries in Table 5-7? The following could be considered:

- When arriving at the site, the operators review the potential hazards associated with the job as follows:
  - Operators slowly rotate 360 degrees and verbally list each of the hazards that are visible at the job site. They can verbally discuss the hazards with each other as they rotate around their work site. If an operator is alone, he or she can comment on each hazard or can enlist the help of a colleague over the radio.
  - Operators repeat the activities above each time they return from a break from the job, e.g., for coffee, for lunch, to retrieve a part or a tool, etc.

The keys to the above procedure are:

- Repetition, so the hazards are recognized over time
- Verbal rehearsal, so the hazards are imprinted in memory

5.3   FACILITY ASSESSMENTS

5.3.1   Preliminary Hazard Analysis

The objective of a Preliminary Hazard Analysis is to identify plant hazards early in a project. Once hazards are identified, the analysis team looks for potential causes for each hazard and makes suggestions for eliminating or mitigating the hazards. The project may be a grassroots addition or a modification to an existing unit. The following material is based on references (Ref. 5-13) and (Ref. 5-14).

The team consists of highly experienced workers from a range of disciplines, as follows:

- Project engineering
- Process engineering
- Operations and maintenance
- Health, safety and environmental advisors

The analysis may be conducted using an adaptation of the "HAZID Checklist"© originally developed by Shell Oil Company and modified in Table 5-8.

Table 5-8.   Hazard identification checklist

Category	Concern	Examples
External and Environmental Hazards		
Natural and environmental hazards	— Climate extremes	— Temperature, waves, wind, dust, flooding, sandstorms, ice, blizzards
	— Lightning	
	— Earthquakes	
	— Erosion	— Ground slide, coastal, riverine

Table 5-8.   Hazard identification checklist (continued)

Category	Concern	Examples
External and Environmental Hazards		
Natural and environmental hazards (continued)	— Subsidence	— Ground structure, foundations, reservoir depletion
Utility systems	— Firewater systems	
Created (man-made) hazards	— Security hazards	— Internal and external security threats
	— Terrorist activity	— Riots, civil disturbance, strikes, military action, political unrest
Effect of the facility on the surroundings	— Geographical infrastructure	— Plant location, plant layout, pipeline routing, area minimization
	— Proximity to population	
	— Adjacent land use	— Crop burning, airfields, accommodation camps
	— Proximity to transport corridors	— Shipping lanes, air routes, roads, etc.
	— Environmental issues	— Previous land use, vulnerable fauna and flora, visual impact
	— Social issues	— Local population, local attitude, social/cultural areas of significance
Infrastructure	— Normal communications	— Road links, air links, water links
	— Communications for contingency planning	— Siren, cellular, 2-way radio
	— Supply support	— Consumables/spares holding

Table 5-8.   Hazard identification checklist (continued)

Category	Concern	Examples
External and Environmental Hazards		
Environmental damage	— Continuous plant discharges to air	— Flares, vents, fugitive emissions, energy efficiency
	— Continuous plant discharges to water	— Target/legislative requirements, drainage facilities, oil/water separation
	— Continuous plant discharges to soil	— Drainage, chemical storage
	— Emergency/upset discharges	— Flares, vents, drainage
	— Contaminated ground	— Previous use or events
	— Facility impact	— Area minimization, pipeline routing, environmental impact assessment
	— Waste disposal options	— Area
	— Timing of construction	— Seasons, periods of environmental significance
Control methods/ philosophy	— Staffing/operations philosophy	— Effect on design, effect on locality (manned, unmanned, visited)
	— Operations concept	— 1-Train, x-trains, simplification
	— Maintenance philosophy	— Plant/train/equipment item, heavy lifting, access, override, bypass, commonality of equipment, transport
	— Control philosophy	— Appropriate technology, (DCS/local panels)

Table 5-8.   Hazard identification checklist (continued)

Category	Concern	Examples
Facility Hazards		
Control methods/ philosophy (complete)	— Staffing levels	— Accommodation, travel, support requirements, consistency with operations and maintenance philosophies
	— Emergency response	— Isolation, ESD philosophy, blowdown, flaring requirements
	— Concurrent operations	— Production, maintenance requirements
	— Start-up, shutdown	— Modular or plant wide
Fire and explosion hazards	— Stored flammables	— Improper storage, operator error, defect, impact, fire (mitigation measures include: substitute non flammable, minimize and separate inventory)
	— Sources of ignition	— Electricity, flares, sparks, hot surfaces (mitigation measures include: identify, remove, separate)
	— Equipment layout	
	— Fire protections and response	
	— Operator protection	

Table 5-8.   Hazard identification checklist (continued)

Category	Concern	Examples
Facility Hazards		
Process hazards	— Inventory	
	— Release of inventory	
	— Overpressure	
	— Over/under temperature	
	— Excess/zero level	
	— Wrong composition/phase	
	— Fuel gas	
	— Heating medium	
	— Diesel fuel	
	— Power supply	
	— Steam	
	— Drains	
	— Inert gas	
	— Waste storage and treatment	
	— Chemical fuel storage	
	— Potable water	
	— Sewerage	

Table 5-8.   Hazard identification checklist (continued)

Category	Concern	Examples
Facility Hazards		
Maintenance hazards	— Access requirements	
	— Override necessity	
	— Bypasses required	
	— Commodity of equipment	
	— Heavy lifting requirements	
	— Transport	
Construction/existing facilities	— Tie-ins (shutdown requirements)	
	— Concurrent operations	
	— Reuse of materials	
	— Common equipment capacity	
	— Interface - shutdown/blowdown/ ESD	
	— Skid dimensions (weight handling equipment (congestion))	
	— Soil contamination (existing facilities)	
	— Mobilization/demobiliz-ation	

The process for conducting the analysis uses a structured brainstorming technique. The flow of the analysis is illustrated in Figure 5-7.

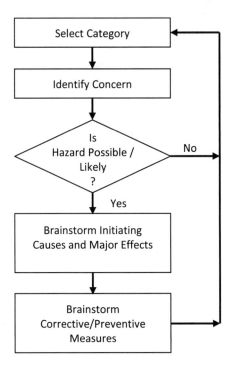

Figure 5-7.   Process flow for Preliminary Hazard Analysis

In preparation for the analysis, the team leader should bring together available documentation, including:

- All previous analyses
- Applicable regulations
- Baseline surveys
- Business philosophy documents
- Development plans
- Lists of products and materials used
- Material Safety Data Sheets
- Plant layouts

Each hazard category from Table 5-8 is considered one at a time using Table 5-9. The team then brainstorms the 'Possible Causes," "Major Effects" and "Safeguards." Once completed, the next hazard category is considered and so on, until each section of Table 5-9 is completed.

Table 5-9.   Preliminary Hazard Analysis

Category	Concern	Possible Cause	Major Effects	Safeguards

The simplified "Preliminary Hazard Analysis" technique described above must not be confused with Process Hazard Analysis (PHA), which is a more rigorous evaluation of process hazards with an operating system.

Could the technique described above be used to identify the corrosion pictured below?

What other technique might be more appropriate?

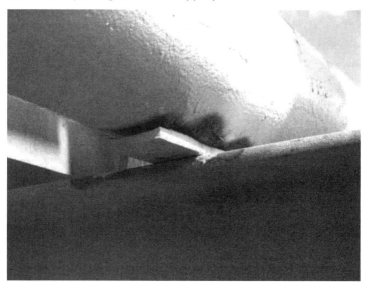

5.3.2   Critical Task Identification Analysis

Critical Task Identification Analysis (CTIA) is a systematic method of identifying critical tasks within a process facility, prioritizing their importance, analyzing those tasks that are considered most critical, and identifying appropriate safeguards to mitigate the risk.   Critical in this context means high risk. The tasks that may have the potential for personal injury, equipment or facility loss, environmental release, community exposure, or business interruption are defined as "critical." These tasks represent a hazard and the goal is to identify the hazards and eliminate/mitigate them before they turn into incidents.

The phases in the CTIA process are presented in Figure 5-8. The following paragraphs describe each phase, record examples of the products of each phase, and provide insights on the most efficient methods to obtain information.

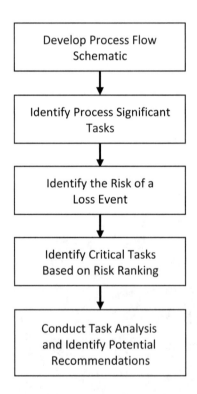

Figure 5-8.   The Critical Task Identification Analysis (CTIA) process

Experience has shown that the analysis is best conducted by a small group of process workers (site team) who, together, have the knowledge necessary to complete the task. The suggested site team consists of:

- Process Technician
- Maintenance Technician
- Facility Engineer
- Health and Safety Specialist

If a human factors specialist is available, their involvement is encouraged.

5.3.2.1 Develop Process Flow Diagram

The analysis starts by specifying the facility processes and equipment and creating a process flow diagram. A simplified process flow diagram is shown in Figure 5-9.

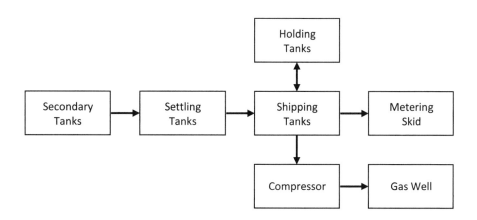

Figure 5-9.   Simplified process flow schematic

5.3.2.2 Identify Process Significant Tasks

The Critical Task Identification process is a 'brainstorming' technique that is designed to encourage the site team to identify the 'critical' tasks that they could be required to perform in each area of the process facility. A critical task could occur during a process upset or other abnormal situation, or it could be a task that is difficult to perform or performed frequently. At each block of the process flow schematic, the team is asked to identify the 'critical' tasks performed by placing them into one of seven different categories including:

- Safety-critical tasks, e.g., responding to a pump seal failure
- Quality-critical tasks, e.g., responding to an out-of-limits lab analysis
- Production-critical tasks, e.g., responding to the loss of an exchanger which reduces output
- Most frequently performed tasks, e.g., filter changes
- Difficult to perform or complex tasks
- Time-critical tasks, e.g., responding to tasks that, in a short period of time, will cause a larger upset (compressor shutdown)
- Environmentally-critical tasks, e.g., responding to an out-of-limit deviation (a high SO_2 reading on a fired heater stack)

INCIDENT HIGHLIGHT: TOXIC RELEASE, 2005

- **Explosion and toxic release**
- **5 fatalities**
- **70 injuries**
- **10,000 evacuated**
- **River contaminated**

Failure to understand and perform safety critical tasks can lead to significant incidents.

As can be seen from the examples in Table 5-10, the same 'critical' tasks could be categorized in different ways. The purpose of the exercise is to identify the major tasks (it is not important which list they appear in). At this point, it may be useful to list the tasks identified in a table such as that provided in Table 5-10.

Table 5-10.   Example critical tasks with descriptions

Process Schematic Section	Task Name	Task Category	Task Description	Risk Ranking (H/M/L)
Upgrader	PSV testing and inspection	Production, Regulatory, Environmental	Test and inspect valves using non-invasive test methods	M
	Tube cleaning	Safety	Remove exchanger tubing, clean in the field and replace	L
Sales Gas	Conduct In-Line Inspection (ILI) of 16" lines	Safety, Environmental, Production	Install device to clean 16-inch gas lines, then retrieve device downstream	H

Task Categories
1. Production critical
2. Safety critical
3. Environmentally critical
4. Frequently performed
5. Difficult to perform
6. Quality critical
7. Time critical

5.3.2.3 Identify the Risk of a Loss Event

No company can afford to spend money and commit resources on issues that don't impact the safety of its workers or might have little benefit to the operation. Thus, it is essential to identify the tasks which, when improved, will improve safety, reduce cost or improve the environment the most. The approach used at this stage is to estimate the risk associated with each task identified in the previous exercise. Many companies have developed a risk matrix to help rank risks to safety and the operation. It is important that the matrix be used for the application for which it is intended (some matrices are intended for higher order risks at a corporate level). The next step in this exercise is to prioritize each 'critical' task in terms of its risk to the facility.

Figure 5-10 illustrates a generic risk matrix that is designed to help workers prioritize tasks. It is presented for illustration purposes only and is not intended for use in formal risk assessments. Experience has shown that a site team, as configured above, can easily categorize tasks into the simplified severity/likelihood risk matrix shown in Figure 5-10.

		Likelihood			
		Frequent	Possible	Rare	Remote
Severity	Major	Very High	Very High	High	Moderate
	Serious	Very High	High	Moderate	Low
	Minor	High	Moderate	Low	Low
	Incidental	Moderate	Low	Low	Low

Figure 5-10.   Example risk matrix

Alternatively, the team can use the risk matrix developed for their company. Tasks that could result in incidents like the one pictured in Figure 5-11 should be ranked with a "HIGH" consequence.

Figure 5-11. Refinery fire

5.3.2.4 Identify the Highest Risk Tasks

The tasks that most require analysis are those with the highest risk in Figure 5-10 (shown in RED). The analysis could also be extended to Medium risk tasks shown by the YELLOW sections and the lower risk tasks shown in the GREEN sections of Figure 5-10.

5.3.2.5 Conduct Task Analysis and Identify Potential Safeguards

The task analysis that is proposed herein is a modification of the process that is described in detail (Ref. 5-12). The analysis process consists of the following two steps.

Step 1: Prepare a Functional Description of Task to be Analyzed

Table 5-11 is completed for each task analyzed, like the task pictured in Figure 5-12. By completing this table, each member of the analysis team gains the same basic level of familiarity with the task. Moreover, the details of the task are discussed, (e.g., the task can be performed 24 hours per day) so the team is now aware of the conditions under which the task can be performed (e.g., poor lighting at night, operators could be fatigued at 3 a.m., etc.). A major requirement at this stage is to supply sketches of the equipment, procedures, layout drawings, or photographs that will assist the team in the analysis.

Figure 5-12.   Cleaning the exterior of process equipment
(See color insert)

Table 5-11.   Functional task description

					Page _____ of _____
Date: _____					Revision: _____
Prepared by: _____					
Job Name: _____					
Job Objective: _____					

Job Function Description

Number of people involved in task	Operating Environment	When is task performed?	Where is the task performed?	Personal protective equipment in addition to normal issue? ☐
	Extreme Temperatures? ☐ High winds? ☐ High vibration? ☐ Poor lighting? ☐ High noise? ☐	Day Only ☐ 24-hour Shift ☐	Indoor ☐ Outdoor ☐	Describe:

Communication	Frequency		Vision
Verbal ☐ Radio ☐ None ☐	Hourly ☐ Daily ☐ Monthly ☐	Yearly ☐ As Needed ☐	Need to View Controls and/or Displays? Describe:
Who with? _____			

Review Checklist

Review Item
Insert sketch, drawing or model of area (attach drawings if applicable)

Step 2: Conduct the Task Analysis

Table 5-12 is used to guide the analysis and record the results. The table consists of the following five columns:

- **Column 1: Potential Human Factors Issues** - Consider this column as a checklist that is designed to help team members consider the human factors issues that need to be considered for each activity.
- **Column 2**: **Activity** - This column is used to list the sequential activities that are performed in the task. Each activity should begin with an action verb, e.g., OPEN a flange, CHECK a display reading, FILL a column.
- **Column 3**: **Potential Hazard** - *What are the hazards that are present for each activity?* For example, wrench slips and injures an operator; line is not purged before it is disconnected. Tools can fall and injure operators below; misreading a gauge can overpressure a vessel.
- **Column 4**: **Ergonomic and Human Factors Concerns** - For each activity, the potential human factors concerns are identified using the job aid in Column 1. From the potential concerns, those that are likely to occur are highlighted.
- **Column 5**: **Potential Hazard Mitigations** - A solution is recommended for each potential hazard that is identified. Solutions usually fall into the following five categories:
  1. Engineering Design: Changes or additions are made to the equipment, facilities, environment or software to mitigate the hazard.
  2. Training: To improve the knowledge and skill of the operator who performs the task.
  3. Job or Task Design: The task conditions are modified. Modifications might include:
     - Changing the times that the task is performed (e.g., day instead of night)
     - Changing the activity sequence
     - Improving procedures, permitting systems, or job aids
     - Reassigning task steps to different work posts
     - Modifying work or shift schedules

4. Selection: A rational strategy would consist of choosing people with capabilities that match the job.
5. Behavior: Behavior is defined as "what people do or say" (Ref. 5-3). Behavior is observable and measurable. It can be affected by engineering design, job design and training. But, there are strategies specifically designed to change behavior. These can include:
   - Creating positive consequences for desired behavior
   - Removing negative consequences for desired behavior
   - Influencing risk perception among workers
   - Creating negative consequences for undesired behavior
   - Providing coaching and corrective feedback for undesired behavior

Which of the jobs pictured below should/could be analyzed using the Critical Task Analysis technique? (See color insert)

Table 5-12. Task analysis worksheet

Date: _____
Prepared by: _____
Job Name: _____
Job Objective:

Page _____ of _____
Revision: _____

Potential Human Factors Issues	Activity	Potential Hazard	Human Factors Issue or Concern	Potential Hazard Solution / Mitigation
Physical Activities *Manual Materials Handling* Hose handling Barrels, boxes Valves height orientation force *Musculoskeletal* hands/wrists upper extremity (head, neck, shoulder) lower extremity (lower back, leg) *Access, walkways, platforms* *Routes: Exit, Entrance, stairs* **Receiving Information** Lighting Noise Interference Display design Labeling Signage Color coding **Processing Information** *Too much information?* *Menu, checklist present?* *Short-term Memory* *Long-term Memory* *Controls/displays as expected?*				

5.4   INCIDENT AND NEAR-MISS REPORTING

Most facilities have robust incident investigation programs that identify root causes and make recommendations to prevent the incident from recurring. Reporting all incidents, regardless of severity, is essential for a successful incident investigation *and hazard identification program.*

One of the opportunities for improvement in an incident investigation program is to enhance or expand near-miss reporting. Near-misses provide an excellent opportunity for hazard identification and mitigation - thus eliminating/mitigating the hazard *before* an incident occurs. A 'near-miss' is defined as *an event or series of events that could have resulted in one or more identified undesirable consequences, but did not.* Traditional near-miss programs in the process industries are conducted to identify potential incident situations and their causes before they occur. Near-misses typically occur much more frequently than incidents, so a robust near-miss reporting program can identify potential incident situations much faster than can incident investigations. However, traditional near-miss programs can fail for a number of reasons including:

- They can be time consuming, especially if operators are asked to participate in an in-depth analysis of their reports
- Operators lose interest in reporting if they sense that their reports are not being considered
- Workers may fear reprimand

Treated properly, near-miss reporting provides a valuable learning tool. Moreover, the reporting system can affect reporting frequency. Paper-based systems require a convenient supply of reporting forms and deposit boxes. Telephone answering systems need to be cleared so they don't fill up. Internet-based systems require convenient access to computers. No matter what reporting system is used, it is important to ensure that the data collected meet the objectives of the near-miss program. For example, if the objective of the near-miss program is to identify hazardous condition causes, it is important for the reporting system to provide:

- Incident categories
- Cause categories
- Location of incident
- How the incident occurred
- How serious could the incident have been

A near-miss format can be used to identify hazards for the purpose of preventing serious incidents, like those shown in Figure 5-13.

Figure 5-13. Tank fire

5.4.1   Hazard Trending and Analysis

Typically, people are often better able to recognize patterns (trends) than to develop concepts from numbers (tables listing process history). Moreover, humans are often able to identify a potential problem from a trend line before many analytical models currently in use (Ref. 5-17). Humans have the ability to filter noise from data and detect trends in them before models can identify a problem. Finally, they have what has been called 'profound' knowledge - the ability to balance the data they see with their experience and intuition (Ref. 5-18). While humans are excellent at analyzing trends, they can't possibly keep up with the large number of variables that have to be monitored. Developing analytical methods that can operate on trend data to provide an early warning can help the operator perform his or her job.

Safety trending typically employs some form of statistical quality control to analyze trends and to advise users when a process is heading out-of-control. A trend is defined as "a statistically significant change in performance measure data which is unlikely to be due to random variation in the process" (Ref. 5-19).

Trending can be applied to hazard recognition. Consider the data in Figure 5-14. A facility has a program to identify and plot leaks from steam lines. Figure 5-14 plots the number of steam leaks that are counted during morning rounds for 20 days.

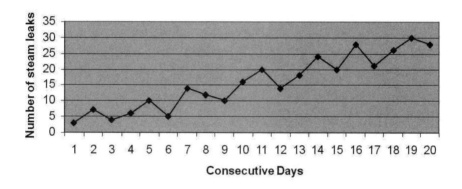

Figure 5-14. Number of steam leaks per day

The number of steam leaks observed is trending upwards at an average increase of about 1.3 leaks per day. Clearly, this is a process that is quickly growing to an unacceptable level and requires a solution.

Now consider Figure 5-15. The steam leaks are not trending upward. The mean number of leaks per day is 20.5 with a standard deviation of 3.5. The question is whether 20 steam leaks per day are acceptable. While every organization plans for zero problems in their operations, it may not be possible.

- Airlines plan for zero lost bags
- Utilities plan for no power outages
- Automobile companies plan for zero warranty claims on their vehicles
- Hospitals plan for no surgical errors

As long as the factors that affect the above outcomes vary, it will not be possible to avoid problems. Nevertheless, the goal is still zero and actions can be taken to drive continuous improvement towards this goal.

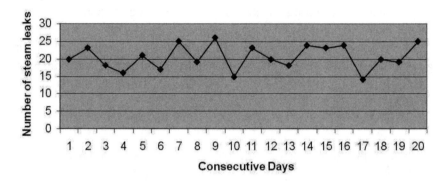

Figure 5-15. Number of steam leaks per day

Trending in the form of statistical quality control charts is used to monitor the improvements in the dependent variable (e.g., steam leaks) and the independent variables that cause the leaks. The result is an outcome that effectively controls risk for an equally acceptable level of cost and effort.

Six sigma is a problem-solving methodology that is used primarily to improve business and organization performance (Ref. 5-20). Six sigma methodologies have been used by many large corporations over the past 20 years, (Ref. 5-21), (Ref. 5-22) and (Ref. 5-23), to save billions of dollars and prevent countless injuries.

5.4.2   Hazard Mapping

Participatory mapping is a technique that is used to identify facility hazards and bodily injuries and illnesses. The technique is best implemented as a group process. In **HAZARD** mapping, workers illustrate their workplace and, collectively, superimpose hazards that occur in their workplace directly on the illustration. The hazards are coded by type (e.g., dust, noise, chemicals) and risk. **BODY** mapping is accomplished by drawing injuries and illnesses directly onto a full body figure (back and front). The concept of body mapping was first introduced to identify cumulative trauma injuries (Ref. 5-24). Participatory mapping has several advantages over other hazard or injury identification processes. First, it is anonymous, since the data is generated by a group of individuals. Second, the data is reliable, since it is developed by a group of individuals who, together, have more complete information of the issues than any one individual. And finally, the process has proven to be a valid representation of actual facility conditions and injury occurrences (Ref. 5-25).

The process of participatory mapping is described in several publications, (Ref. 5-25), (Ref. 5-26), and (Ref. 5-27). For hazard mapping, groups of six to ten workers from the same work area begin by sketching their work place or by developing a process flow schematic of their site (Figure 5-16). The group then identifies a list of hazards (using, for example, Figure 5-16) that are present onsite and create icons for them. The next step is to locate the hazard icons on the facility sketch or process schematic at the place that each hazard occurs in the facility. The symbols can be additionally coded for high, medium or low risk. The completed sketch can be used for hazard mitigation exercises and discussions with management.

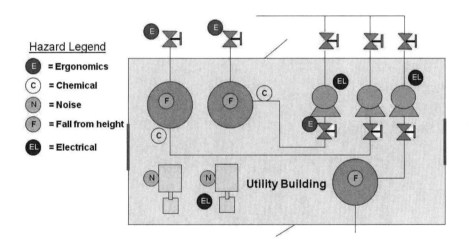

Figure 5-16.   Hazard mapping example

Body mapping is conducted similarly. Small groups of workers identify injuries and illnesses that occur in the work place and graphically illustrate their body location (e.g., carpal tunnel, lower back injuries, etc.) on full body illustrations (Figure 5-17). Again, the injuries can be coded for severity or by frequency of occurrence.

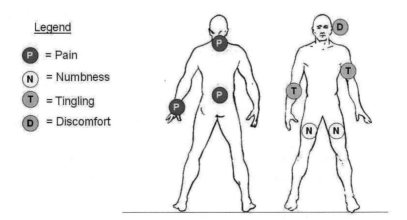

Figure 5-17.   Body mapping example

5.5   HAZARD IDENTIFICATION AND ANALYSIS TRAINING

Training in Hazard Identification and Analysis (Figure 5-18) can be one of the most effective ways to identify and eliminate or mitigate the effects of local plant hazards.

Figure 5-18.   Conducting training on hazard identification and analysis

The objective of every training program should be to maximize skills and knowledge (Ref. 5-12). Hazard training programs can be designed not only to improve worker knowledge on hazards and provide the skills to address them, but also to use the training time to identify hazards that are specific to a plant. Referring to the hazard management process logo that introduces each chapter, the training program could revolve around four major hazard categories:

- Process
- Physical
- Equipment
- Electrical

Examples are provided in Table 5-8.

At the end of each training session on one of the above categories, the training class could be broken into groups to brainstorm the specific hazards in their plant.

Group brainstorming is very effective since the result is a list of local, plant hazards that would not have been identified without worker input.

5.6   REFERENCES

5-1.   Stanton, N.A., Baber, C., and Young, M.S. "Observation". Chapter 28 in N. Stanton, A. Hodge, K. Brookhuis, E. Salas and H. Hendrick (eds) *Handbook of Human Factors and Ergonomic Methods.* CRC Press, New York, NY. 2004.

5-2.   Bennett, J.D. "Loss Prevention System." ExxonMobil Supported Behavior-Based Safety System; © James D. Bennett. 1997.

5-3.   Geller, E. S. "Working Safe: How to Help People Actively Care for Health and Safety." Chilton Book Company, Radnor, PA. 1996.

5-4.   McSween, T.E. "The Values-Based Safety Process: Improving Your Safety Culture with a Behavioral Approach." Van Nostrand Reinhold, New York, NY. 1995.

5-5.   Krause, T.R. "Worker-Driven Systems for Safe Behavior: Integrating Behavioral and Statistical Methodologies." Van Nostrand Reinhold, New York, NY. 1995.

5-6.   Crawley, F., and Tyler, B. "Hazard Identification Methods." European Process Safety Center, Institution of Chemical Engineers, Rugby, Warwickshire, UK. 2003.

5-7.   Kemmlert, K. "PLIBEL – The Method Assigned For Identification of Ergonomic Hazards." Chapter 3 in N. Stanton, A. Hodge, K. Brookhuis, E. Salas and H. Hendrick (eds) *Handbook of Human Factors and Ergonomic Methods.* CRC Press, New York, NY. 2004.

5-8.   U.S. Department of the Navy. "Manual Materials Handling, Checklist and Solutions Guide." Naval Facilities Engineering Command.

5-9.   Oregon Occupational Safety and Health Administration. "Conducting a Job Hazard Analysis (JHA)." OR-OSHA 103

5-10.   Occupational Safety and Health Administration. "Job Hazard Analysis" U.S. Occupational Safety and Health Administration 3071. 2002.

5-11.   Roughton, J.E. and Crutchfield, N., "Job Hazard Analysis: A Guide for Compliance and Beyond." Elsevier, New York, NY. 2008.

5-12.   Attwood, D.A, Deeb, J.M., and Danz-Reece, M.E.D. "Ergonomic Solutions for the Process Industries."

5-13.   Center for Chemical Process Safety. "Guidelines for Hazard Evaluation Procedures: Third Edition." American Institute of Chemical Engineers, New York. 2008.

5-14.   Shell Oil Company. "HAZID Process: EP 95-0312"

5-15.   Van de Schaaf, T.W. and Wright, L.B. "Systems for Near-miss Reporting and Analysis." Chapter 33 in J.R. Wilson and N.Corlett (eds) *Evaluation of Human Work, Third Edition."* Taylor and Francis, CRC Press, New York, NY. 2005.

5-16.   Department of Energy. "Root Cause Analysis Guidance Document." U.S. Department of Energy, Office of Nuclear Energy, Washington, D.C. 1992.

5-17.   Lu, S. and Huang, B. "Intelligent Process Trend Recognition Fault Diagnosis and Industrial Application." In *Computational Intelligence."* Vol 4114. Springer-Verlag (publishers), Berlin. 2006.

5-18.   Deming, W.E. "The New Economics for Industry, Government, Education: 2nd Edition." MIT (publishers), MA. 1994.

5-19.   Prevette, S.S. "Making Business Decisions Using Trend Information." Presentation to the 1997 U.S. Department of Energy, TRADE Conference, December 2. 1997.

5-20.   Gygi, C., De Carlo, N., and Williams, B. "Six Sigma for Dummies." Wiley Publishing, Hoboken, NJ. 2005.

5-21.	George, M., Rowlands, D. and Kastle, B. "What Is Lean Six Sigma?" McGraw-Hill, New York, NY. 2004.

5-22.	Summers, D. C. S. "Six Sigma: Basic Tools and Techniques." Pearson (Prentice Hall), Upper Saddle River, NJ. 2007.

5-23.	Wedgewood, I.D. "Lean Sigma: A Practitioner's Guide" Prentice Hall, Upper Saddle River, NJ. 2006.

5-24.	Corlett, E., and Bishop, R. "A Technique for Assessing Postural Discomfort." Ergonomics, 19: 175-182. 1976.

5-25.	Keith, M.M., Brophy, J.T., Kirby, P., and Rosskarn, E. "Barefoot Research: A Worker's Manual for Organizing Work Security." International Labour Organization, Geneva, Switzerland. 2002.

5-26.	CUPE. "Enough Overwork: Taking Action on Workload." Canadian Union of Public Workers, Ottawa, Ontario. 2002.

5-27.	Keith, M.M. and Brophy, J.T. "Participatory Mapping of Occupational Hazards and Disease among Asbestos-Exposed Workers from a Foundry and Insulation Complex in Canada." International Journal of Occupational and Environmental Health. 10, 144-153. 2004.

6

MAKE RISK-BASED DECISIONS

After identifying and evaluating hazards, the next step is to make decisions based on the magnitude of the risk. Figure 6-1 illustrates this step in relation to the hazard management process.

REMEMBER:

**RISK = THE SEVERITY OF THE HAZARD X THE LIKELIHOOD
THAT THE HAZARD WILL CAUSE HARM**

Figure 6-1. Hazard management process

Hazard ranking involves evaluating the risk associated with the hazard. To accomplish this, it is important to understand two things about the hazard:

- How severe are the consequences (without any safeguards or layers of protection)?
- How likely are the consequences to occur?

A **hazard** is a condition (often a substance or object) with the potential to cause harm. **Risk** is a combination of the severity of that harm and the likelihood that it will occur. For example:

- The hazard is flammable material. The risk is that a pump seal failure will result in a release of flammable material, fire and explosion and personnel injury.
- The hazard is electricity. The risk is that a worker might be electrocuted when exposed to inadequately insulated electrical wires.
- The hazard is caustic soda. The risk is that a worker might suffer a chemical burn if his or her skin is directly exposed during batch mixing operations.

It is important to thoroughly understand a hazard before attempting to quantify or rank it. Simple techniques based on a series of questions can help achieve such an understanding. Depending on the results of the following questions, it may be possible to pursue further information by adding additional questions.

Simple Hazard Evaluation Questionnaire

1. What is the hazard?
2. Is it harmful in its current state or is another event or condition required to initiate harm?
3. What type of harm can it cause?
4. What safeguards or layers of protection exist that could reduce the harm resulting from the hazard?
5. Are the existing safeguards adequate?
6. What alternatives exist for dealing with this hazard?

Hazard evaluation needs to consider that certain hazards might have a different effect on individual workers. The factors that could contribute to such anomalies include the health and physical condition of a worker, his or her position relative to the hazard, worker experience (tendency to react or respond), weather, visibility, and any physical obstructions. Any hazard evaluation should be based on the cautious or conservative side and consider the worst likely effect on workers.

It's important to recognize that there are unique hazards associated with the start-up and shutdown of industrial processes and equipment. Every time raw materials are introduced into a system or the operating conditions are brought up to normal state, the system undergoes transition. This change in status can also place additional demands on people. There is sometimes a tendency to take shortcuts and make compromises during start-up since conditions are of short duration and may be deemed only temporary. This must not be allowed to happen. Start-up is a critical period in the life cycle of any facility; it sets the stage for a run-cycle that may last several months or years. Vigilance during facility commissioning and start-up is especially important. Field personnel should keep a watchful eye out for unusual conditions and be prepared to deal with hazards even if the start-up has to be delayed. Failure to identify start-up hazards and manage the associated risks can result in significant incidents, like the one described below.

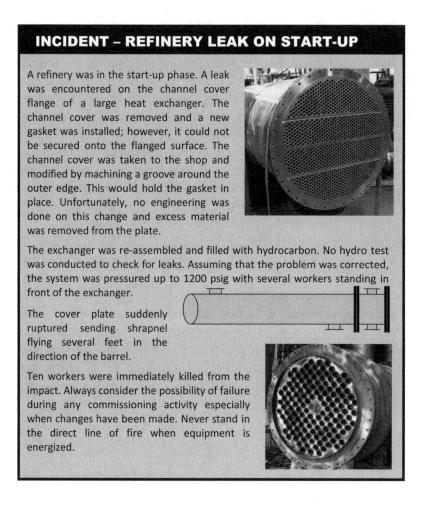

INCIDENT – REFINERY LEAK ON START-UP

A refinery was in the start-up phase. A leak was encountered on the channel cover flange of a large heat exchanger. The channel cover was removed and a new gasket was installed; however, it could not be secured onto the flanged surface. The channel cover was taken to the shop and modified by machining a groove around the outer edge. This would hold the gasket in place. Unfortunately, no engineering was done on this change and excess material was removed from the plate.

The exchanger was re-assembled and filled with hydrocarbon. No hydro test was conducted to check for leaks. Assuming that the problem was corrected, the system was pressured up to 1200 psig with several workers standing in front of the exchanger.

The cover plate suddenly ruptured sending shrapnel flying several feet in the direction of the barrel.

Ten workers were immediately killed from the impact. Always consider the possibility of failure during any commissioning activity especially when changes have been made. Never stand in the direct line of fire when equipment is energized.

What were the hazards and how should they have been addressed?

How could this entire incident have been avoided?

Any system under high pressure is vulnerable to leakage or mechanical failure. Such occurrences can propel fluids or mechanical components towards workers with great force. Workers should never stand in the direct line-of-fire of mechanical components or system openings that could experience high pressure.

6.1   HAZARD RANKING

As previously suggested, there may be more than one hazard associated with a work activity.

A simple hazard ranking process can enable workers to make better choices when safety problems arise in the field. While hazard ranking is not a substitute for formal risk analysis, it can quickly warn of the need to stop work or to choose an alternate strategy. Hazard ranking is conducted on the basis of the type of harm that a hazard might inflict. The principal challenge is to recognize any safeguards or layers of protection that might be present to prevent the exposure or to minimize the effects.

The simple thermometer diagram illustrates the concept of hazard ranking (Figure 6-2).

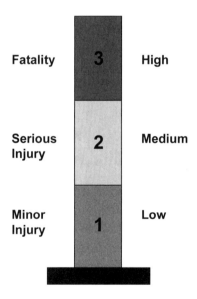

Figure 6-2.   Concept of hazard ranking

Using the hazard ranking process described above and illustrated in Figure 6-2, how would you rank the hazards depicted in Figure 6-3?

Figure 6-3.   Rank the hazards

Again, it is important to define the industrial setting in which these exposures might be encountered. The reader is urged to apply his or her own experience and imagination.

6.2   UNDERSTANDING RISK

Understanding the risk (the product of severity and likelihood) posed by a hazard is essential towards establishing an effective hazard management strategy. High risk hazards (ones that could result in serious consequences and are more likely to occur) must be given the highest priority. A system to prioritize and rank risks can help to ensure hazard management is approached in the most efficient way. Such a system should include:

- A consistent method for hazard evaluation
- A means to focus resources (time, resources and money) to eliminate or mitigate the highest risk hazards first
- A tool for employees to better understand and evaluate hazards

There are risks associated with almost any activity. Activities such as processing and transporting flammable, toxic or corrosive materials; mining and transporting materials; transporting people; and disposing of waste have associated hazards and risks, but these may be necessary to support a country's economy.

There is often more than one hazard involved in a single work activity. A risk assessment may involve identifying and evaluating several hazards and risks, selecting the hazards to address as well as deciding how to reduce the associated risks to a lower level.

To attempt to eliminate all risk would not be possible or economical. Yet, every reasonable effort should be taken to reduce risks to a minimum practical level. Such a concept was introduced into UK safety legislation in the 1970's. Referred to as the ALARP (As Low as Reasonably Practicable) principle, it mandates aggressive risk reduction measures in critical operating systems to as low a level as practical. The target level of low risk defined by ALARP may vary from company to company and will depend on technical knowledge, experience and availability of skilled resources.   Some companies and jurisdictions stipulate a target value for ALARP.   The remaining risk is termed "residual risk".

Companies may be expected to defend their risk mitigation strategies against common industry practices.

Workers should be encouraged to reduce risks to as low a level as practical given available resources and other constraints. Ultimately, a supervisor or team leader may be required to defend the hazard management strategy adopted by the workers.

Is this risk acceptable?

6.3   RISK RANKING

Risk matrices provide a visual platform for gauging and communicating the magnitude of a risk. Proper use of a risk matrix can help to ensure that a risk is perceived and dealt with in a consistent manner. Risk matrices reflect a corporation's risk management practices and, in particular, the willingness to deal with a given level of risk. While the use of a risk matrix is common across industry, there is not one particular size or format that is typical.

The example in Figure 6-2 portrayed hazard assessment as a simplified one-dimensional process. However, sub-categories of Severity and Likelihood are required to evaluate risk and focus risk decisions more precisely. In addition, if there is more than one hazard identified, more defined sub-divisions of Severity and Likelihood help prioritize hazards – thus providing the company a tool for determining which hazards to address sooner and which ones can be deferred until later. Figure 6-4 illustrates a simplified risk matrix, with sub-categories for Severity and Likelihood.

		Likelihood		
		Frequent	Occasional	Unlikely
Severity	High	3	3	2
	Medium	3	2	1
	Low	2	1	1

Figure 6-4.   Simplified matrix for field hazard/risk assessment
and ranking

In the above risk matrix, a risk assessment score of "3" is intended to trigger immediate remedial action.   A risk assessment score of "2" is intended to trigger a remedial action at an appropriate time (with possible need for prompt interim action).   A risk assessment score of "1" suggests that the hazard is more moderate and some discretion might be applied in dealing with it.

Any risk matrix should adapt itself readily to the type of business or industry. Whether it is used to make corporate risk decisions or to set field safety priorities will determine the ranges on the severity and likelihood scales and the allocation of risks into the high, medium and low categories. Each company or organization will need to establish such a basis. Regardless of the size and format adopted, there is one noteworthy precaution: a procedure or guideline should be followed to assign the hazard and risk rankings.   This procedure should require developing the hazard scenario *before* determining the likelihood. The use of a risk matrix by an untrained worker could lead to faulty and subjective conclusions.

In this discussion of using risk to drive action, it is important to note that any issue that is identified as a hazard entails some degree of risk.   As such, even if the risk is deemed low and can be managed, it may warrant a periodic revisit to assure that it has not increased in severity or likelihood.

Hazards can be identified on a more formal and structured basis. An example of this might be a team facility inspection guided with a checklist. In such a case, multiple hazards may be identified. The risk ranking of those hazards and implications of actions to take based on their relative risk rankings suggest the use of more formal approaches. These are similar to risk ranking approaches used in Process Hazard Analyses and other formal hazard identification methods.

A basic tool used to rank the severity and likelihood of groups of identified hazards is an expanded risk matrix (Figure 6-5). This example is a two dimensional matrix (Severity and Likelihood) with four sub-divisions of Severity and Likelihood. It is important to recognize that each block in the matrix typically represents a range of consequence and likelihood. The matrix shown in Figure 6-5 is termed a 4 X 4 and is a common size that is used in formal hazard assessments, such as Process Hazards Analysis. This risk ranking matrix assigns values for the severity of the consequences and the likelihood of the occurrence to determine a risk value. This risk value is then used to assign a priority to the scenario that needs to be addressed. Risk ranking matrices vary in size, but typically range from a 4 x 4 to a 6 x 6, depending on their intended application.

| | | Likelihood | | | |
		Frequent	Possible	Rare	Remote
Severity	Major	Very High	Very High	High	Moderate
	Serious	Very High	High	Moderate	Low
	Minor	High	Moderate	Low	Low
	Incidental	Moderate	Low	Low	Low

Figure 6-5.   Simplified 4 X 4 risk ranking matrix

To use any risk matrix, the severity and likelihood categories should be defined. Furthermore, it is important that the matrix be used for its intended application or work environment. Large corporations sometimes have separate matrices for safety, environmental, and economic risks. If the matrix is intended to support corporate risk decisions, the threshold level of risk acceptance may be higher than that of a small plant or facility. It is important to check the intended purpose of a matrix before it is used.

In using this matrix, the recommended approach is to estimate the severity of the outcome due to exposure to the hazard. Then, the likelihood (frequency) of the hazard reaching that level of severity is estimated. Using those two determinations, a "ranking" of the risk can be read from the matrix using the intersection of the severity and likelihood.

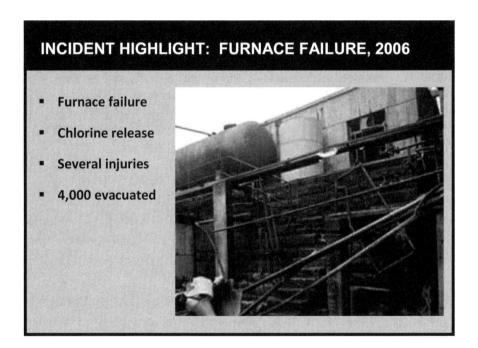

INCIDENT HIGHLIGHT: FURNACE FAILURE, 2006

- **Furnace failure**
- **Chlorine release**
- **Several injuries**
- **4,000 evacuated**

6.3.1   Severity

In the example matrix shown in Figure 6-5, potential severities of the consequences are shown in Table 6-1.

Table 6-1.   Range of severities*

	Injury	Environmental	Business Interruption/ Asset Damage	Reputation
Major	One or more fatalities	Offsite release with short or long-term damage	Destruction of process unit, significant long-term downtime	Extended national media coverage
Serious	Serious, multiple injuries	Offsite release with immediate remediation	Significant equipment damage, extended downtime	National media coverage - limited
Minor	Single injury, not severe	Onsite contained release	Equipment damage, short-term downtime	Extended local coverage
Incidental	Minor or no injury	No impact	Limited or no downtime	Minor story in local media

Varies by Company

Many companies also consider the severity of the environmental impact and the business interruption/asset damage consequences and include these on their matrices. Some companies also include loss of reputation.

The priority of corrective actions should reflect the highest risk that a company is trying to manage. However, the actions taken to mitigate other risks to tolerable levels must also reduce the safety risks to tolerable levels.

Regardless of what size the risk ranking matrix is, the value being assigned for severity (the severity ranking) must be based ONLY on the severity of the ultimate consequences. No consideration is given for safeguards or layers of protection when ranking the severity.

How would you rank the severity of a hazard that could result in an explosion, like the explosion that devastated a sugar refinery, as shown in Figure 6-6?

Figure 6-6. Sugar refinery explosion *(See color insert)*

Accumulation of combustible dust on flat surfaces led to the explosion pictured above.
This type of hazard is easily recognized and mitigated.
When dust hazards are ignored, their consequences may be catastrophic.

6.3.2   Hierarchy of Safeguards or Layers of Protection

Processing facilities require a variety of strategies and approaches to keep the process risk at tolerable levels.   When evaluating and designing a protection system, it is important to address the concept of "common mode failure," where safeguards or layers of protection are susceptible to failure from a common cause. For example, an electrical power loss, utility failure, weather, instrument miscalibration, or process contaminants or conditions (e.g., foaming) all may affect multiple controls, instrumentation, or other safeguards or layers of protection. Having all critical alarms and interlocks in the same basic process control system makes them all susceptible to failure when that system fails or loses power.   Similarly, different sensors using the same technology to measure normal level and to activate a high level interlock on a vessel are susceptible to failure by common process conditions (e.g., fouling), where both may fail to work properly. By using diverse protection measures, the chance of common failure of safeguard systems can be minimized and preparedness for loss events can be enhanced. Finally, it is critical to assure that protective systems are independent of the initiating cause of the risk situation being evaluated.

6.3.2.1 Inherently Safer Design

Inherently safer design is a concept, an approach to safety that focuses on eliminating or reducing the hazards associated with a set of conditions. While implementing inherently safer design concepts will move a process in the direction of reduced risk, it will not remove all risk. Inherently safer design consists of:

- **Minimize:** Use smaller quantities of hazardous substances (also called *Intensification*).
- **Substitute:** Replace a material with a less hazardous substance.
- **Moderate:** Use less hazardous conditions, a less hazardous form of a material, or facilities that minimize the impact of a release of hazardous material or energy (also called *Attenuation* and *Limitation of Effects*).
- **Simplify:** Design facilities which eliminate unnecessary complexity and make operating errors less likely, and which are forgiving of errors that are made (also called *Error Tolerance*).

When possible, elimination of the hazard is the ultimate goal of hazard management. For example, many facilities who traditionally used one-ton cylinders of chlorine for water treatment at cooling towers have switched to sodium hypochlorite. Chlorine is immediately dangerous to life and health (IDLH) at 10 ppm. The Emergency Response Planning Guideline Level 2 (ERPG-2) is 3 ppm. A release of one ton of chlorine will produce a vapor cloud that may travel up to 3 miles in concentrations that are hazardous to the community and require evacuation and/or shelter-in-place. Exposure to sodium hypochlorite can cause topical burns to exposed workers. Switching from chlorine to sodium hypochlorite eliminates the vapor cloud hazard and provides substantial risk reduction.

6.3.2.2 Engineering Controls

To manage and control risks, the industry has incorporated many engineered safety systems into our processing facilities. When hazard elimination is not possible, the next best options are engineered solutions. These solutions can be either passive or active.

A passive approach uses safety solutions that do not require an action to be taken. Examples of passive solutions are curbs around equipment, fire or blast walls, or robust design of pressure vessels to handle the expected overpressure. Another example is an enclosure around a piece of noisy equipment to reduce the noise exposure to personnel (Figure 6-7).

Figure 6-7.   Noise enclosure

Active approaches require that a physical action occur in response to a process event. An example is a Burner Management System (BMS) that does not allow lighting of a furnace until the firebox has been purged. Figure 6-8 illustrates a BMS panel. Another example is a Safety Instrumented System (SIS) that brings a process to a safe state when a process deviation exceeds the safe operating conditions. Active solutions are generally less reliable than passive solutions.

Figure 6-8. Burner Management System panel

6.3.2.3 Administrative Controls

Administrative controls provide another safeguard or layer of protection, but should not be relied on in lieu of practical engineered controls. Administrative approaches that require human action can increase the likelihood of human error.   Examples of administrative controls are procedural checks, operator actions in response to an alarm, and emergency response following a loss event.

Many companies use administrative controls for operations that require personnel to follow specific steps (e.g., checklist). This could be for either operational or maintenance procedures.

Other examples of administrative control include lockout/tagout (Figure 6-9), and a car-seal on the inlet of a pressure relief device that is car-sealed open (Figure 6-10).

An administrative procedure can be used to control the amount of material in storage so that in the event of a release there is not a sufficient quantity of material to cause an offsite issue with the community.

Figure 6-9.   Lockout/tagout example

Figure 6-10. Carseal

6.3.3   Likelihood

After assigning a severity value, the next step is to assess how likely (or probable) the consequences are to occur. In the 4 x 4 matrix depicted in Figure 6-5, the likelihood of the consequences can be qualitatively described as:

- **Frequent:**   The scenario has occurred at the facility or is reasonably likely to occur at any time.
- **Possible:** The scenario is likely to occur at this facility.
- **Rare:** The scenario has occurred at a similar facility and may reasonably occur at this facility during its anticipated lifecycle.
- **Remote:** Given current safeguards or layers of protection, the scenario is not likely to occur at this facility.

When determining likelihood, a facility's past history with a particular hazard, process or category of equipment should be considered. Safeguards or layers of protection should also be taken into consideration to avoid overstating the risk. When safeguards or layers of protection are not taken into consideration when determining likelihood, the resultant risk value is known as the unmitigated risk. This value can be useful in conveying just how significant a risk is if safeguards or layers of protection are not present and maintained. This can help maintain and communicate a sense of vulnerability to employees. When safeguards or layers of protection are taken into consideration, the resultant risk value is known as the mitigated risk. Priority of safeguards or layers of protection is given for engineering controls, followed by administrative controls, safeguards or layers of protection that mitigate the severity after an incident has occurred (e.g., emergency response); and finally personal protective equipment, respectively. Example safeguards or layers of protection are illustrated in Figure 6-11. To simplify further discussion, the remainder of this chapter focuses on mitigated risk.

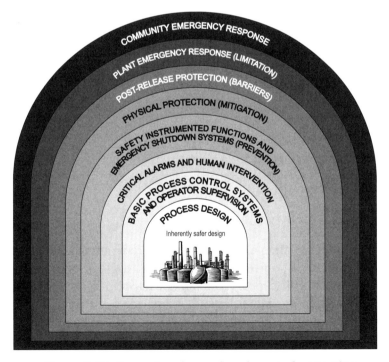

Figure 6-11. Example safeguards or layers of protection

6.3.4   Risk Ranking

The resultant risk described in the matrix in Figure 6-5 has the following values:

- **Very High:** Additional safeguards or layers of protection are required to eliminate or mitigate the hazard. More sophisticated techniques may be needed to analyze the risk in order to make decisions. In some cases, this risk reduction must be promptly implemented before the job task can continue.
- **High:** Additional safeguards or layers of protection are required to eliminate or mitigate the hazard.
- **Moderate:** Additional safeguards or layers of protection should be considered and prioritized with other site risks.
- **Low:** Additional safeguards or layers of protection are typically not required.

Examples of "Very High" risk scoring may include:

- **Asset Damage/Financial Loss:** Significant leak of flammable material with potential vapor cloud explosion.
- **Community/Environmental Impact:** Equipment conditions that suggest a high potential for a major leak (e.g., catastrophic vessel failure resulting in large release of toxic gas cloud).
- **Injury/Illness:** Personnel working inside a confined space with potential for asphyxiation.

Examples of "High" risk scoring may include:

- **Asset Damage/Financial Loss**: Crane lifting heavy load drops material onto pipe bridge, destroying process equipment. Note: If process equipment is pressurized and contains hazardous materials, the risk may be deemed high.
- **Injury/Illness:** Personnel working at heights without fall protection, resulting in permanent impact injuries.

Examples of "Moderate" risk scoring may include:

- **Asset Damage/Financial Loss**: Piping leak releases hazardous material to soil or water.
- **Community/Environmental Impact:** Piping leaks to soil or water.
- **Injury/Illness:** Personnel not wearing proper personal protective equipment, resulting in contact with acid requiring medical treatment.

Examples of "Low" risk scoring may include:

- **Asset Damage/Financial Loss**: Seepage of water pump seal, resulting in business interruption.
- **Community/Environmental Impact:** Slight odor in process area, resulting in nuisance disturbance to neighbors/community.
- **Injury/Illness:** Maintenance materials left on the job, resulting in a slip/trip/fall and first-aid case.

As stated previously, the objective is to reduce the hazard's risk to its lowest practical level. The examples above are for illustrative purposes only and may not represent the risk within a "real" facility.

6.3.5   Example 1 - Flammable/Explosive Hazard

Using the first example of a hazard at the beginning of the chapter:

- The hazard is a flammable liquid. The risk is that a pump seal failure will result in a release of flammable material, fire and explosion and personnel injury.

6.3.5.1 Case 1

Let us assume that there is a highly flammable liquid in the line. A seal failure could result in a release that forms a vapor cloud. The vapor cloud could travel to a pump motor or vehicle on the road, which provides an ignition source and could result in a fire and explosion.

Additional information about the scenario is:

- The pump is equipped with a single seal. The pump has experienced a history of pump seal failures.
- There is an operator working in the area.

Using Figure 6-5, the risk ranking of this hazard scenario would be:

- **Severity = Major**. *Justification*: An operator working in the area where a fire and explosion occurs could sustain fatal injuries.
- **Likelihood = Frequent**. *Justification*: The history of the pump seal failures combined with the lack of safeguards or layers of protection results in a frequent likelihood of occurrence.
- **Risk = Very High**.   Additional safeguards or layers of protection are required. Based on the history of the pump seal failure, many facilities would implement additional safeguards or layers of protection before the pump is returned to service.

Risk ranking of this hazard scenario indicates that hazard mitigation should be given a very high priority. The decision may be made to not return the pump to service until additional protection layers are provided.

6.3.5.2 Case 2

Let us assume that there is a highly flammable liquid in the line.   A seal failure could result in a release that forms a vapor cloud. The vapor cloud could travel to a pump motor or car on the road, which provides an ignition source and could result in a fire and explosion. Additional information about the scenario is:

- The pump is equipped with a tandem seal which vents to a safe location.
- There is an operator working in the area.

Using Figure 6-5, the risk ranking of this hazard scenario would be:

- **Severity = Major.** *Justification:* An operator working in the area where a fire and explosion occurs could sustain fatal injuries.
- **Likelihood = Seldom.** *Justification:* The tandem seal venting to a safe location reduces the likelihood that a seal leak that results in a release will occur.
- **Risk = High.** An additional safeguard or layer of protection or safeguard is required.

Do you think the facility described in the incident below understood their hazards, the safeguards or layers of protection to manage the hazards, and their resulting risk?

INCIDENT – TOXIC RELEASE AT PULP MILL

Shortly after midnight, in the fall of 1994, the top section of a large wooden stave vat of bleached pulp collapsed at a large pulp mill. The force of the impact ruptured two adjacent tanks of chlorine dioxide, a highly hazardous chemical used to bleach pulp.

This material is very toxic and corrosive and can cause death and serious injury even at low concentrations. Because it was late at night, there were few people on site and no fatalities or serious injuries were sustained. While the condition of the vat was not readily obvious to the workers, the proximity of the large vat to other equipment was a concern and should have received some follow-up attention. There were no secondary containment berms or means to divert the spill from sensitive equipment nearby. The viscous contents of the large vat caused an external failure of the steel tanks containing the high hazard material. Emergency planning must recognize and address such high hazard scenarios resulting from tight equipment spacing or inappropriate layout.

6.3.5.3 Case 3

Let us assume that there is a highly flammable liquid in the line.   A seal failure could result in a release that forms a vapor cloud. The vapor cloud could travel to a pump motor or car on the road, which provides an ignition source and could result in a fire and explosion. Additional information about the scenario is:

- The pump is equipped with a tandem seal which vents to a safe location.
- There is an operator working in the area.
- There is a critical instrument with an alarm on the seal pot.

Using Figure 6-5, the risk ranking of this hazard scenario would be:

- **Severity = Major.** *Justification:* An operator working in the area where a fire and explosion occurs could sustain fatal injuries.
- **Likelihood = Unlikely.** *Justification:* The critical instrument with an alarm on the seal pot will alert the operator to pump problems before the secondary seal fails. The tandem seal venting to a safe location reduces the likelihood that a seal leak resulting in a release will occur.
- **Risk = Moderate.**   Additional safeguards or layers of protection may not be required.

6.3.6   Example 2 - Flammable Hazard

Changing the example of a hazard at the beginning of the chapter:

- The hazard is a combustible liquid in the line with a high flashpoint. The risk is that a pump seal failure will result in a release of flammable material causing fire and personnel injury.

Let us assume that there is a combustible liquid with a high flashpoint in the pipe. A seal failure could result in a release that could ignite and result in a fire. Additional information about the scenario is:

- The pump is equipped with a single seal. The pump has experienced a history of pump seal failures at this facility.
- There is an operator working in the area.

Using Figure 6-5, the risk ranking of this hazard scenario would be:

- **Severity = Serious.** *Justification:* An operator working in the area where a pump seal fire occurs could sustain serious injuries.
- **Likelihood = Frequent.** *Justification:* The history of the pump seal failures combined with the lack of protection layers results in a frequent likelihood of occurrence.
- **Risk = Very High.**   Additional safeguards or layers of protection are required. Based on the history of the pump seal failure, many facilities would implement additional safeguards or layers of protection before the pump is returned to service.

The addition of tandem seals which vent to a safe location reduces the likelihood to occasional, and the resultant risk to moderate. The tandem seals can also have a seal pot with an alarm that indicates loss of barrier fluid, which will further reduce the likelihood to unlikely.

Pump seal fires can escalate to major facility fires (Figure 6-12). The safeguards or layers of protection discussed above help reduce the risk and prevent a pump seal fire from escalating and causing significant damage to the facility and potential serious consequences to personnel.

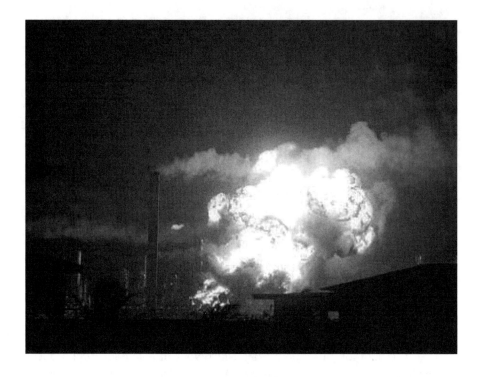

Figure 6-12. Pump seal fires can escalate!

How could hazard identification and making risk-based decisions have prevented the incident described below?

INCIDENT – ROOF COLLAPSE ON PROCESS EQUIPMENT

An ore treatment plant handled large quantities of water, steam and mineral tailings. Because the plant was located in a colder climate, all process equipment was enclosed in a large steel clad building. An ore treatment plant handled large quantities of water, steam and mineral tailings. Because the plant was located in a colder climate, all process equipment was enclosed in a large steel clad building.

Several holding tanks and process vessels were vented to atmosphere within the building. The roof of the building was located 60 ft above the floor level and was supported by conventional open web steel joists similar to those used in many warehouses. During cooler weather conditions, steam condensed on the cold roof panels and support joists; this caused the joists to rust. Operators were well aware of this problem but they did not respond, believing that their primary focus was the operation of the process. One winter, following a heavy snowfall, a large section of the roof collapsed onto the process equipment below. Fortunately, no one was killed or injured. Mechanical damage was extensive and the plant was required to shut down for costly repairs. Hazard identification must extend beyond the primary business of the facility or workplace. Changes or deteriorating conditions in the supporting infrastructure, such as roads and buildings, that could affect people or the process should be reported and addressed. A mechanical failure caused by process or environmental conditions can have a significant impact.

Could this incident have been prevented?

Could a different workplace culture have prevented the incident?

Could the build-up of snow have been observed by someone and should it have been cleared?

Should someone have inspected the roof beams for corrosion and deterioration and highlighted the concern? Even if you are not an expert, you should be able to identify that something is simply not right.

6.3.7   More Detailed Matrices

The use of an expanded matrix provides greater resolution by providing better breakdown of the options one can chose (Figure 6-13). However, where there is uncertainty in judgment or complex problems need to be analyzed; more sophisticated techniques (i.e., Layer of Protection Analysis (LOPA) or Quantitative Risk Assessment (QRA)) are typically employed to provide the necessary resolution and precision.

6.3.8   Similarities Between More Sophisticated Process Hazard Evaluation Techniques

There are several other sophisticated methods to analyze and rank hazards and risks. Operations and maintenance workers will occasionally be required to participate in these risk processes – **hazard identification is the important first step.**

Although the objective of this book does not include exploring other hazard evaluation techniques, CCPS has published several other books and guidelines, including:

- *Guideline for Hazard Evaluation Procedures*, Third Edition, American Institute of Chemical Engineers, Center for Chemical Process Safety, New York, New York, 2008.
- *Layer of Protection Analysis, Simplified Process Risk Assessment,* American Institute of Chemical Engineers, Center for Chemical Process Safety, New York, New York, 2001.

The type of hazard evaluation study is determined by the type and complexity of the process. Some companies have established protocols for determining the appropriate techniques, which can include:

- Checklist
- What-If?
- Hazard and Operability (HAZOP) Study
- Failure Mode and Effects Analysis (FMEA)
- Fault Tree Analysis

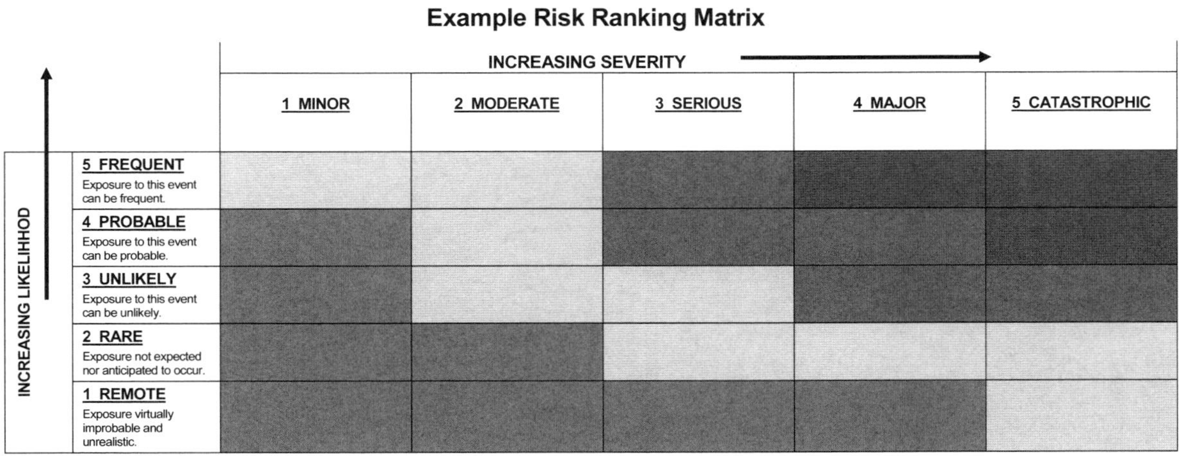

Example Risk Ranking Matrix

INCREASING SEVERITY →

	1 MINOR	2 MODERATE	3 SERIOUS	4 MAJOR	5 CATASTROPHIC
5 FREQUENT Exposure to this event can be frequent.					
4 PROBABLE Exposure to this event can be probable.					
3 UNLIKELY Exposure to this event can be unlikely.					
2 RARE Exposure not expected nor anticipated to occur.					
1 REMOTE Exposure virtually improbable and unrealistic:					

← INCREASING LIKELIHHOD

Green: **Low** – *Low Risk.* No mitigation required.

Yellow: **Medium** – *Medium Risk with Controls Verified.* Verify proper risk assessment. Risk may be acceptable as is with existing controls, safeguards, procedures, etc.; are verified as functional and in place. Higher range categories may require additional action to reduce risk. **ALARP** should be evaluated, as necessary.

Orange: **Significant** – *Significant Risk.* Manage risk utilizing prevention and/or mitigation *with priority.* Promote issue to appropriate management level with commensurate risk assessment detail.

Red: **High** – *High Risk.* Mange risk utilizing prevention and/or mitigation *with highest priority.* Promote issue to appropriate management level with commensurate risk assessment detail.

A risk matrix is a tool used for comparative ranking of risks. Risk matrices are company specific and should have a sound technical basis. They should be used in conjunction with an established risk assessment process. The risk matrix should reflect the company's risk policies. Each company should establish a defensible basis before adopting a risk matrix. The use of a corporate risk matrix to make important safety and risk decisions in most organizations should be referred to a qualified risk specialist.

Figure 6-13. Example 5 x 5 risk matrix

When using any of these techniques, it is important to ensure the participation of a multi-disciplinary team, including representatives from maintenance and field operations.

It is interesting to note the similarities in the way hazards are evaluated using different techniques. The Hazard and Operability (HAZOP) technique, often used to conduct Process Hazard Analyses, would most likely have identified a similar hazard scenario as the one discussed previously:

The hazard is a Class I Flammable material. The risk is that a pump seal failure could result in a release of flammable material, fire and explosion and personnel injury.

6.4   REFERENCES

6-1.   Center for Chemical Process Safety. *Guidelines for Hazard Evaluation Procedures*. Center for Chemical Process Safety of the American Institute of Chemical Engineers. New York, 1985.

7

FOLLOW-UP AND CALL TO ACTION

After understanding how to recognize, evaluate, and eliminate/mitigate hazards, it's time to implement your hazard management program. As discussed in Chapter 1, there are two essential steps to establish an effective hazard management program (Figure 7-1):

1. Management commitment to providing the resources, empowering employees, and measuring and managing the process.
2. Employee ownership of the program to ensure that it is effectively implemented and maintained.

Figure 7-1.   Hazard management process

7.1   SAFETY CULTURE

Safety culture has been defined as, "the combination of group values and behaviors that determine the manner in which safety is managed" (Ref. 7-1).

"Companies have found that if safety and health values are not consistently shared at all levels of management and among all workers, any gains that result for declaring safety and health excellence a "priority" are likely to be short-lived" (Ref. 7-2).

A strong safety culture develops as a group identifies certain attitudes and behaviors that provide common benefit to its members, in this case, attitudes and behaviors that support the goal of safer process operations. As the group reinforces such attitudes and behaviors, and becomes accustomed to their benefits, these attitudes and behaviors become integrated into the group's value system. In an especially sound culture, deeply held values are reflected in the group's actions, and newcomers are expected to endorse these values in order to remain part of the group.

A company's process safety culture directly impacts its hazard management program. Process safety management system failures can often be linked to weak safety culture. Accordingly, enlightened organizations are increasingly seeking to identify and address such cultural root causes of process safety performance problems. A good safety culture is reflected in the hazard management process by:

- Integrating safe operations into the company's core values.
- Focusing on proactive identification of hazards, clear understanding of risk, and effective risk reduction/mitigation measures
- Providing resources proportional to the perceived risk(s) it seeks to control.
- Placing an emphasis on learning from past experience in order to prevent future problems.
- Striving to continuously improve performance.
- Empowering employees to be involved in identifying hazards and deciding how they should be addressed.

Additional information on culture can be found in *Guidelines for Risk Based Process Safety*, Chapter 3 (Ref. 7-3).

7.2   MANAGEMENT COMMITMENT

Why Get Management Commitment? The simple answer is that success is dependent on management commitment. Time invested in gaining management's commitment pays off in tangible support, which not only facilitates - but also expedites a long-term process. Commitment, as the term is used here, refers to explicit, concrete actions, not merely to rhetoric.

Management commitment can be defined as direct participation by all levels of management (including foremen and front-line supervisors) of an organization in the specific and critically important aspects of the program. In the hazard management process, it includes:

- Developing and sustaining a culture that embraces hazard identification
- Identifying, understanding, and complying with codes, standards, regulations, and laws
- Establishing and continually enhancing organizational competence
- Soliciting input from and consulting with all workers and contractors

---

A workforce that is convinced the organization fully supports safety as a core value will tend to do the right things, in the right way, at the right times - even when no one else is looking (Ref. 7-3).

7.3   EMPLOYEE OWNERSHIP

To be committed to their job, employees must have ownership of the work. To have ownership of the work, employees must be able to influence what goes on in their workplace. And to influence the workplace, employees must be heard and reasonably answered by their supervisors (Ref. 7-4).

Important drivers in employee commitment include:

- Trust in senior leadership
- Opportunity to make a difference in the workplace
- Chance to use skills on the job
- Job security
- Absence of work-related stress
- Honesty and integrity of their company's business conduct

When employees take ownership of their work activities, hazard management and safety performance can be dramatically improved.

Before these job tasks were started, how would you have communicated the hazards to the people performing the work? (See color insert)

7.4   IMPLEMENT AN EFFECTIVE HAZARD MANAGEMENT PROGRAM

Key elements of an effective Hazard Management Program are described in the following sections.

7.4.1   Written Procedures and Training

After reviewing the variety of techniques that can be used to identify workplace hazards (Chapter 5), select an appropriate combination of techniques that are appropriate for identifying hazards at your worksite. (Remember - these techniques can be used to identify different types of hazards).

The next step is to develop written procedures and guidance for conducting the techniques you selected and train all workers in the techniques and the objectives of a hazard management program.

7.4.2   Resolving Recommendations for Risk Reduction

A program's effectiveness will be directly impacted by timely closeout of recommendations. When workers see changes being made as a result of their efforts, they become more involved and have more ownership for the program - they become "champions" for hazard management.

7.4.2.1   Hazards Requiring Immediate Action

There are some hazards that require immediate action - there simply isn't time to do an "evaluation" or "look at options" for risk reduction. These types of hazards could result in serious injury and have limited protection layers (on the risk ranking matrices in Chapter 6, they would have the highest risk ranking). When the stakes are this high, the job task or process needs to be immediately stopped and a supervisor notified immediately. Workers must be empowered to stop work under these conditions. An example would be a flange leak or a high level leak in a tank.

7.4.2.2   Managing Recommendations

Recommendations resulting from workplace hazard management programs should be tracked to closure, responsibility assigned, documented, and communicated to affected workers.   Chapter 6 discussed the hierarchy for recommendations:

- Inherently Safer Design
- Engineered Controls
- Administrative Controls

What types of hazard controls would you apply to the following job tasks?

How can management commitment and employee ownership affect these controls? (See color insert)

If a recommendation will not be closed-out immediately or in a timely manner, communicating this status to workers is essential so that they are assured that management has taken it seriously and that the issue remains on management's radar.

7.4.3   Concepts to Strengthen Protective Systems

Training, behavior modification and workers selection are actions that can be used to strengthen/enhance a safeguard - **but these are not safeguards by themselves**. Because training is an easy (and low cost) recommendation, workers often rely on training for addressing a hazard, when additional engineered controls should be implemented. (It is important to note that formal risk assessment techniques do not take risk reduction credit for training (Ref. 7-5)).

7.4.3.1   Training

Training is essential to a safe work environment and should not be a substitute for a hazard that can be "engineered-out." Training of operations workers on the proper response to alarms and the actions they should take is critical to reducing the potential for hazardous events. More importantly, training all workers to be observant while they are in areas where hazards may exist is the best resource for indentifying hazards.

For example: Maintenance workers, while working in a unit to repair a pump, notice a vibration on an adjacent pump and notify operations. This notification allowed operations to shut down the pump before the seal fails, preventing a release of hazardous material.

7.4.3.2   Behavior Modification

The most effective types of behavior modification result from positive motivation - when a worker understands the reason for changing the way they perform their job. Simply telling them that an approach is "bad" will not be as effective in changing their behavior as providing them an explanation of why a different way is safer or more effective (Figure 7-2).

"It might help Rover's behavior if you said he needed improvement instead of calling him a bad dog."

Figure 7-2.   Ineffective behavior modification

Behavior modification may not typically be thought of as a hazard protection approach, however it can be beneficial. Workers learn certain methods for operating a unit or repairing a pump. They may not have been taught the best or the safest method for that particular operation, but the problem is that they have been doing it that way for many years. Simply teaching them the correct method will not work, they will revert back to the old way in a short period of time. *What is needed to change their behavior?*

An example is a maintenance technician who has been repairing a pump and aligning the seal using a method he learned several years ago; however, with a recent change in seal design, his method of alignment of the seal results in seal failure after a short period of use. The technician attended training on the new seal alignment, but did not recognize that the method was different. In order to break his previous behavior pattern, a concerted effort to provide quality assurance review of his work until he demonstrates understanding is required.

7.4.3.3   Personnel Selection

Today's workforce is more educated and more in tune to thinking outside-the-box. The personnel selection process needs to identify the necessary skill set for operations and maintenance for performing their assigned job function.   One element of the skill set is the ability to recognize when something is not right. This does not mean that they may know what is wrong, but that they sense (through the basic human senses discussed in Chapter 3) that there is a problem.

7.5   HAZARD COMMUNICATION

Once a hazard has been identified and it is determined that it cannot be eliminated, then what? Communication of the hazard to those who could be impacted is critical to avoiding potential harm. Communication can take many forms:

- Center for Chemical Process Safety (CCPS) Process Safety Beacon
- Communication at tailgate or safety meeting
- Communication up the line and communication from the top down
- Communication between individual workers
- Process safety documentation, such as periodic, 5-year risk assessments
- Safety bulletins
- Shift turnover logs
- Signs and barricades
- Training programs
- Work permits

Some companies have a formal risk register and have already highlighted the significant events and provided response training.

The simplest form of communication is individual discussion between workers; however, the disadvantage to this method is not everyone may be told.   Hence, a system for communication of hazards should be available. Also, communication of the hazards needs to include the safety effects (and environmental and other effects, if appropriate) associated with the hazard. The hazard may be a caustic material. The effect is severe burns and possible blindness if a person is exposed.

Hazards that are short term or may only exist during certain job activities can be communicated through work permits and tailgate or safety meetings. These methods should be used for hazards that are temporary or job related. An example could relate to change-out of equipment at high elevations, so the hazard would be potential for falling objects or workers falling. This hazard would only exist immediately before and after the equipment change. If the hazard will exist for more than a day or two, another approach may be more desirable.

Shift turnover logs are effective in providing details of a hazard for multiple shifts and when the hazard may be present for longer periods of time. An example might be when a compressor is making an unusual noise and the next several shifts are asked to monitor the noise for changes and notify engineering of the change.

There are some hazards that cannot be eliminated and are always present. An example is the use of caustic in the process to remove impurities in a product. The caustic is required to be used, and the hazards should be covered through training programs, Material Safety Data Sheets, and cautions in operating and maintenance procedures. Procedures should also include consequences and appropriate warnings of deviating from the prescribed operation. The scenarios associated with exposure to this material should be periodically reviewed and updated.

In some cases, a hazard may be known, but only at lower levels within an organization. In this case, the hazard needs to be communicated up the line so that management can make a determination as to whether the risk is within the company's tolerance level. Also, the management of a facility must be aware of field hazards and risks so that they can develop realistic business plans. It may happen that the hazard has not been fully understood at low levels; hence, the risk is also not understood.

The reverse could also occur. A new hazard could be introduced through a change made at higher levels in the organization, and the resulting risk needs to be communicated to those affected by the change. This should be communicated through the facility's Management of Change program. A proper hazard review or evaluation is critical when any change is made to existing plants, including equipment, procedural, software, workers, etc. Hazards often creep into an operating facility over time as changes occur if proper management of change review does not occur. Lack of proper hazard identification, analysis, and mitigation through management of change was determined to be the root cause in several major chemical accidents investigated by the U.S. Chemical Safety and Hazard Investigation Board (Ref. 7-6).

7.6   CALL TO ACTION

It's time to break the cycle of incidents that occur in the process industry - and establishing a strong workplace hazard management program is an essential step in this process.

Hazard elimination isn't always possible because the raw materials we use and the products we make have hazardous properties - making hazard management even more important. *(See color insert)*

7.7 REFERENCES

7-1 "Turning the Titanic – Three Case Histories in Cultural Change." Jones, David. CCPS International Conference and Workshop, Toronto. 2001.

7-2 "The Report of... The BP U.S. Refineries Independent Safety Review Panel (Baker Panel Report)." 2007.

7-3 "Guidelines for Risk Based Process Safety." Center for Chemical Process Safety of the American Institute of Chemical Engineers. NY, NY. 2007.

7-4 Management Issues - the heart of a changing workplace. "Listening, the key to employee commitment." http://www.management-issues.com. 2006.

7-5 "Layer of Protection Analysis." Center for Chemical Process Safety of the American Institute of Chemical Engineers. NY, NY. 2001.

7-6 "Management of Change." US Chemical Safety and Hazard Investigation Board. Washington D.C. 2001.

8

LEARNING AND CONTINUOUS IMPROVEMENT

"What has happened before will happen again.

What has been done will be done again.

There is nothing new under the sun."

Ecclesiastes 1:9

(See color insert)

A practical approach to hazard identification for operations and maintenance workers always involves learning from past incidents and applying those lessons learned to achieve continuous improvement. Throughout the previous chapters, incident examples have been provided to reveal the importance if identifying and managing hazards. It's always easier to learn from past mistakes, the key is not to repeat them.

8.1   CASE STUDY - OIL REFINERY FIRE, 2007

A massive fire injured four workers and caused the total shutdown and evacuation of an oil refinery in February of 2007. The incident likely occurred after water leaked through a valve, froze, and cracked an out-of-service section of piping, causing a release of high-pressure liquid propane (Ref. 8-1).

The fire occurred in the refinery's propane de-asphalting unit, which uses high-pressure propane as a solvent to separate gas oil from asphalt; gas oil is used as a feedstock in other gasoline-producing refinery processes. The propane leaked from an ice-damaged piping elbow that is believed to have been out-of-service since the early 1990s. Unknown to refinery personnel, a weld rod was left in the pipe work and wedged under the gate of a manual valve above the piping elbow, allowing liquid to flow through the valve. Piping above the valve contained liquid propane at high pressure, and small amounts of water were entrained in the propane.

Over time, water seeped past the leaking valve and built up inside the low point of the piping elbow. A period of cold weather in early February 2007 likely caused the water to freeze, expand, and crack the piping. On February 16, the daytime temperature increased and the ice began to melt. At 2:09 p.m. high-pressure liquid propane flowed through the leaking valve and was released through the fractured elbow. Investigators estimated that propane escaped from the pipe at an initial rate of 4,500 pounds per minute, quickly creating a huge flammable vapor cloud, which drifted toward a boiler house where it is believed it contacted an ignition source.

There was no way to shut off the supply of fuel, because the refinery had not installed remotely operable shutoff valves. The growing fire caused the failure of a pipe flange on a large extractor tower filled with propane, igniting a powerful jet fire that was aimed directly at a major pipe bridge carrying liquid products throughout the refinery. Because the nearby pipe bridge supports were not fireproofed, they quickly collapsed, severing process pipes and adding more fuel to the fire.

The fire also caused the release of an estimated 5,300 pounds of toxic chlorine from three one-ton cylinders stored 100 feet from the fire. The chlorine, used to disinfect cooling water, could have posed a serious threat to emergency responders had they not already been evacuated. In addition, the fire threatened a large spherical tank that contained up to 151,000 gallons of highly flammable liquid butane. As a result of the growing fire, the valves controlling a water deluge system designed to cool the sphere became inaccessible to operators and could not be opened.

An investigation concluded the root causes of the accident were that the refinery did not have an effective program to identify and freeze-protect piping and equipment that was out-of-service or infrequently used; that the refinery did not apply the company's policies on emergency isolation valves to control fires; and that current industry and company standards do not recommend sufficient fireproofing of structural steel against jet fires.

This incident resulted in:
- Four Injured
- Refinery Evacuated
- Extended Shutdown

Key Issues:
- Freeze Protection of Dead-Legs
- Emergency Isolation of Equipment
- Fireproofing of Support Steel
- Fire Protection of High pressure LPG Service
- Chlorine Release

The U.S. Chemical Safety and Hazard Investigation Board (CSB) made recommendations to the American Petroleum Institute (API), a leading oil industry trade association that develops safety practices that are widely followed in the U.S. and overseas (Ref. 8-1). The CSB is an independent federal agency charged with investigating industrial chemical accidents. The CSB called on the API to develop a new recommended practice for freeze-protection of refinery equipment and to improve existing practices related to fireproofing, emergency isolation valves, and water deluge systems.

8.2   IMPORTANCE OF MANAGING CHANGE

Consistent and effective management of change (MOC) is one of the most important and difficult activities to implement in a company (Ref. 8-2). Managing change is important because uncontrolled changes can directly cause or lead to catastrophic events. Formal MOC systems include administrative procedures for the review and approval of changes before they are made. This process helps ensure the continued safe and reliable operation of facilities.

These historically significant incidents indicate the continued difficulty of managing change safely:

- **1974:** An explosion occurred at a chemical plant close to the village of Flixborough, England, on June 1, 1974. 28 people were killed and 36 were seriously injured. The process involved oxidation of cyclohexane with air in a series of six reactors to produce a mixture of cyclohexanol and cyclohexanone. Two months prior to the explosion, a crack was discovered in the No. 5 Reactor. It was decided to install a temporary 50 cm (20 inch) diameter pipe to bypass the leaking reactor to allow continued operation of the plant while repairs were made. On June 1, the temporary bypass pipe ruptured. Within a minute, about 40 tons of cyclohexane leaked from the pipe and formed a vapor cloud which exploded, completely destroying the plant. Around 1,800 buildings within a mile radius of the site were damaged. No review or analysis of the change (installing the temporary bypass line) was conducted.

INCIDENT – CHEMICAL PLANT EXPLOSION

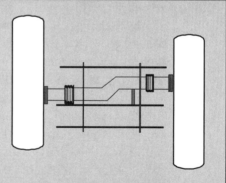

A chemical plant experienced a devastating explosion and fire in 1974 that destroyed the plant, killed 28 workers and injured several dozen others. The plant produced a nylon intermediate from a flammable hydrocarbon stream that passed through a series of high pressure cylindrical reactors.

When a crack was discovered on one reactor, that reactor was removed from service and bypassed using a temporary 20 inch diameter pipe. This pipe was not engineered and was supported by light weight scaffolding. Furthermore, the piping was constructed using miter joints and expansion bellows, neither of which was approved for high pressure and high temperature service. Several days into the run the pipe ruptured releasing flammable hydrocarbon that immediately exploded and burned. The unusual configuration of the piping should have flagged a serious concern to experienced workers. If it looks unsafe, it probably is!

- **1984:** The Bhopal incident was an industrial disaster that took place at a Union Carbide subsidiary pesticide plant in the city of Bhopal, India. On December 3, 1984, the plant released 42 tons of methyl isocyanate (MIC) gas, exposing more than 500,000 people to toxic gases. The incident was a catastrophe for Bhopal with an estimated 3,000 fatalities, 100,000 injuries, and significant damage to livestock and crops. The long-term health effects from such an incident are difficult to evaluate: the International Medical Commission on Bhopal estimated that as of 1994 upwards of 50,000 people remained partially or totally disabled. Among the many root causes of this incident, the failure to manage change at the facility was a significant contributor as equipment and safety systems were taken out-of-service.
- **1998:** On April 8, 1998, an explosion and fire occurred during the production of dye at a chemical plant in Paterson, New Jersey. The explosion and fire were the consequence of a runaway reaction, which overpressured a 2000-gallon chemical vessel and released flammable material that ignited. Nine employees were injured. Although there were several key findings, the CSB's Investigation Report (Ref. 8-3) found that the facility did not use its MOC procedures to review the safety of changes in equipment size and changes in batch size, resulting in increased difficulty of controlling heat output - which ultimately led to a runaway reaction.
- **2005:** On December 11, 2005, an early morning explosion and fire at a fuel depot in Britain experienced an explosion and fire that destroyed several nearby buildings and injured 43 workers. Approximately 2000 local residents were evacuated and a major highway was closed. The fire took 4 days to extinguish and a total of 20 tanks were destroyed. A plume of black smoke hung over the region for several days. Had the incident occurred during daylight hours several lives would have been lost.

This incident was the result of a petrol tank being overfilled after an instrument test procedure. Following the test, the level transmitter was not re-commissioned, a procedural deviation. The rising level in the tank was not detected till the tank overflowed into the common diked area surrounding several tanks. As the petrol spilled over the sides of the tank, some of it vaporized and spread beyond the diked area, finding a source of ignition. The force of the blast was magnified by congestion within the buildings and around the tanks.

Critical test procedures require a high level of human attention. Given reduced workforce staffing during the night shift, one might question the wisdom of filling tanks while an instrument test is taking place. Furthermore, the placement of several tanks within a common diked area places additional challenges on emergency planning. The tank layout should have flagged the possibility of a fully involved tank fire to experienced workers.

In 2008, the Center for Chemical Process Safety (CCPS), released a new publication, *Guidelines for Management of Change for Process Safety*, which is a valuable resource for developing and implementing change management programs (Ref. 8-2).

"Those who cannot remember the past are condemned to repeat it."

George Santayana

Coking Oven Incident

A coking oven was taken out of service for maintenance. The contents of the oven were dumped and the oven was purged with air. When the temperature within the oven had dropped to 100°F and there was little indication of particulates in the air, two contract workers entered the oven through the dump chute located at the bottom of the oven. Several minutes later a large lump of hot material weighing approximately 4 tons dropped from the wall of the oven and buried the two workers. Both workers were killed instantly. An investigation revealed that the material was still extremely hot and had not been adequately scoured from the walls of the oven. It is important to visually observe for signs of unstable foreign material within any vessel prior to entering any equipment.

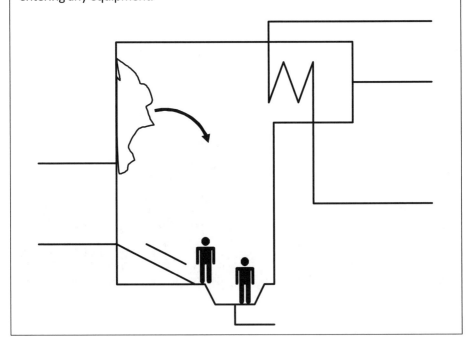

8.3   PUBLISHED ACCIDENT DATABASES AND RESOURCES

While there are a few published accident databases, another way to "learn from someone else's experience" is to network with your peers. Attend conferences where incidents are discussed, but not necessarily published. Establish an internal company network between world-wide locations. Table 8-1 provides information on publically available sources.

Table 8-1.   Accident databases and sources

Resource	What it Contains	Where to Find It
Process Safety Incident Database. Center for Chemical Process Safety of the American Institute of Chemical Engineers.	The Center for Chemical Process Safety developed the Process Safety Incident Database to collect, track and share important process safety incidents and experiences among project participants.	http://www.psidnet.com/ (requires membership)
Process Safety Beacon. Center for Chemical Process Safety of the American Institute of Chemical Engineers.	Monthly flyer that captures current incidents and lessons learned. Examples are contained on the companion CD.	http://www.aiche.org/CCPS/Publications/ Beacon
European Gas Pipeline Incident Datagroup (EGIG)	While the database is not publically available, downloadable reports are available on the website.	http://www.egig.nl/
Incidents that Define Process Safety. Center for Chemical Process Safety Concept Book. (Ref. 8-4)	A summary of incidents that provide valuable "lessons learned" and the areas of process safety where the failures occurred.	http://www.aiche.org/ccps/
U.S. Chemical Safety and Hazard Investigation Board (CSB) website.	Incident investigation reports, incident reports, videos of incidents and investigation results.	http://www.csb.gov/

Table 8-1. Accident databases and sources (continued)

Resource	What it Contains	Where to Find It
Major Accident Reporting System (MARS)	The Major Accident Reporting System (MARS) is a distributed information network, consisting of 15 local databases on a MS-Windows platform in each Member State of the European Union and a central UNIX-based analysis system at the European Commission's Joint Research Centre in Ispra (MAHB) that allows complex text retrieval and pattern analysis.	http://mahbsrv.jrc.it/mars/default.html

8.4   REVITALIZING LESSONS LEARNED

How can we keep reminding ourselves of the past?

How can we keep the lessons learned from incidents in the forefront of our minds and actions?

How can we keep from repeating past events?

The challenge is how to effectively learn and create an environment that encourages learning from both successes and failures.   To maintain a heightened sense of importance and avoid becoming "numb" to the lessons incidents have taught us, there are several things we can do at our facilities to emphasize their importance:

- Establish management commitment to maintaining lessons learned by having upper management involved in sharing lessons from incidents and near misses.
- Provide resources for employees to access industry-wide incident information.
- Provide networking opportunities at conferences or other meetings where incidents and near misses are discussed.

- Encourage employees to share their lessons learned with co-workers, other facilities within your organization, or at conferences and outside meetings.

Remember that different facilities have more things in common than things that are different.   There are significant learning benefits associated with accident case studies at other facilities.

8.5   TRANSFER OF KNOWLEDGE

As companies prepare for the potential mass departure of valuable staff, they are looking toward preserving the knowledge of experienced employees. The process industries have recently experienced staff reductions - leaving the remaining employees with more responsibility, along with significant changes in processes and technology - dramatically increasing their workload.   Managing and retaining their knowledge becomes challenging and extremely vital to maintaining corporate memory.

Based on a recent study, growth in the volume of information available and rapid technological progress has forced most people into a state of information overload (Ref 8-6). "This has left organizations scrambling to create systems for acquiring, retaining, and accessing an overwhelming volume of data.   Added to this is the demand for highly specialized knowledge that is often difficult to find and retain. Knowledge management is one method for ensuring that years of accumulated wisdom do not leave the organization once the employee retires or moves on. The challenge is to create an atmosphere that fosters knowledge sharing, while simultaneously underscoring that transferring knowledge is a way for employees to leave a legacy that will ultimately help the organization long after they leave."

The process of establishing a formal system to manage transfer of knowledge involves these key areas:

- Identifying and collecting information
- Storing information
- Transferring information
- Managing the process

8.5.1   Identifying and Collecting Information

The first step is to identify the information to be collected. This can be accomplished by auditing the job tasks and systems at the facility and establishing an "inventory" of the information that the company wants to capture.   Interviews with subject matter experts (SMEs) can be a valuable source of information.   Observing, interviewing and documenting information from SMEs can help establish best practices, improve procedures and help maintain corporate memory.

8.5.2   Storing Information

Once the information is captured in various forms, the company must decide where the information is best stored so that it is easily used by the employees who need it. Some companies retain information in document management systems or databases.

Refinery Fire 1994

A large integrated oil refinery sustained a lightning strike during an electrical storm. A small fire occurred in one of the main process units while several other units were forced to shutdown due to a power outage. This created a situation whereby several units discharged excess hydrocarbons to the relief and blowdown system. Despite these difficult conditions, the refinery continued to operate the catalytic cracking unit which produces gasoline.

After several hours, the catalytic cracking unit also experienced upset conditions. Attempts to stabilize the unit were unsuccessful. Several instrument failures led to a high level condition in one of the main process columns. Had proper computer diagnostics been applied the upset might have been avoided. Finally, the unit discharged a large quantity of liquid hydrocarbon to the blowdown header that was already in a flooded and partially blocked condition. The corroded blowdown line quickly ruptured resulting in a large fire that injured several workers and destroyed much of the plant.

Operating under upset conditions represents a significant hazard to people, processes and equipment. After a brief period, if stability cannot be achieved in a operation, consideration should be given to shutting down and re-starting using established operating procedures.

8.5.3   Transferring Information

In order to be effective, this information that has been identified, collected and stored must be transferred to the appropriate employees. This can be accomplished through:

- On-the-Job Training
- Mentoring
- Documenting Best Practices in Written Procedures and Training
- Meetings
- Sharing Lessons Learned

8.5.4   Managing the Process

Companies that effectively manage knowledge claim higher rates of productivity. Maintaining corporate memory helps:

- Improve decision making
- Increase efficiency of procedures
- Reduce re-work
- Reduce incidents
- Improve performance

The result is reduced cost of operations and improved performance. Managing transfer of knowledge is especially helpful for large companies where multiple locations and groups prevent workers from knowing and benefiting from the work of others.

8.5.5   Applying What You've Learned

Although training and the transfer of knowledge are critical steps to learning, the only way to achieve continuous improvement without experiencing an incident is to properly apply past lessons learned to current and new processes and procedures. Very few processes are truly unique, however recognizing the similarities and applying the appropriate safeguards takes time and training to be effective.

Gas Plant Explosion 1998

A large gas plant complex sustained a major explosion and fire that killed two workers, injured several others and destroyed a major portion of the plant. A local gas shortage resulted in significant business penalties for the surrounding communities.

For several years leading up to the incident the plant had experienced line blockage problems resulting from hydrate formation in lines. These problems were particularly prevalent during periods of high gas demand. Line restrictions led to surges in pressure and level within several process vessels making it difficult to establish steady state conditions. As a result, the control room panels were often indicating a state of alarm with several lights either on or flashing. This situation became "normal" and as a result no formal reporting or investigation of such incidents took place.

On the day of the incident, a high demand for gas created upset conditions. A critical control loop was taken out of service for maintenance and this caused a flow interruption in a circuit that normally passed light hydrocarbon. This upset the unit heat balance and caused a downstream heat exchanger to become extremely cold. When the flow was manually restored by the field crew a few hours later, thermal shock caused the exchanger to rupture, releasing a large quantity of flammable hydrocarbon. This hydrocarbon exploded and burned leading to the consequences cited earlier.

Control panel lights in a state of alarm should not be viewed as normal. Every alarm signal should trigger a response action whether manual or automatic. If alarm signals become commonplace or accepted as the norm, operators should view this as a high hazard with potential consequences. Be sure to report and investigate any alarm conditions that are not reasonable and which cannot be explained.

8.6   LEARNING FROM INCIDENTS

If the previous sections haven't emphasized the importance of learning from our past, perhaps the next incident examples will:

- In June 2003, about 8,000 pounds of vinyl chloride monomer (VCM) were released at Chemical Plant A when an operator cleaning a reactor mistakenly opened the bottom valve and blind flange on an operating reactor. In this incident, the VCM dispersed without igniting.   Since this incident, all reactor drain valves at the facility are locked when not in use, and only supervisors have access to keys.

- In February 2004, at Chemical Plant B (owned by the same company as Chemical Plant A) an operator mistakenly transferred the contents of an operating reactor to a stripper tank. Consequently, pressure in the stripper tank exceeded its relief device setpoint and VCM released through the vent pipe to the atmosphere. The operator used an emergency air hose to bypass the reactor bottom valve interlock without supervisor authorization.   As a result of the incident, operators were retrained on emergency reactor transfer procedures, including the requirements for supervisory approval prior to the use of the emergency air hose.   A recommendation was made by the facility to redesign the interlock bypass to prevent unauthorized use and established a deadline for completion by April 1, 2004.   The design changes had not been completed by the time of the next incident on April 23, 2004.

- On April 23, 2004, five workers were fatally injured and two others were seriously injured when an explosion occurred in a polyvinyl chloride (PVC) production unit at Chemical Plant B. The explosion forced a community evacuation and lighted fires that burned for several days at the plant.   The CSB concluded that the accident occurred when an operator overrode a critical valve safety interlock on a pressurized vessel making polyvinyl chloride (Ref. 8-7).

The CSB found (Ref. 8-7) that both the current and past owners of the facility, were aware of the possibility of serious consequences of an inadvertent release of chemicals from an operating PVC reactor. But the investigation determined that the measures both companies took were insufficient to prevent human error or minimize its consequences.

Former CSB Chairman Carolyn W. Merritt said, "People do make mistakes. And that is why it is all the more important for chemical plants to design systems that take into account the possibility of such errors." Ms. Merritt continued, "This accident occurred because the companies involved did not look closely enough at the potential for catastrophic consequences resulting from human error" (Ref. 8-7).

"If history repeats itself, and the unexpected always happens, how incapable must Man be of learning from experience."

George Bernard Shaw
(See color insert)

8.7   REFERENCES

8-1.   U.S. Chemical Safety and Hazard Investigation Board. "Investigation Report, LPG Fire at Valero - McKee Refinery." Report No. 2007-05-I-TX. 2008.

8-2.   Center for Chemical Process Safety. "Guidelines for Management of Change for Process Safety." Center for Chemical Process Safety of the American Institute of Chemical Engineers. A John Wiley & Sons Publication. Hoboken, New Jersey. 2008.

8-3.   U.S. Chemical Safety and Hazard Investigation Board. "Investigation Report, Chemical Manufacturing Incident." Report No. 1998-06-I-NJ. 1998.

8-4.   U.S. Chemical Safety and Hazard Investigation Board. "Investigation Report, Refinery Explosion and Fire." Report No. 2005-04-I-TX. 2007.

8-5.   Center for Chemical Process Safety. "Incidents that Define Process Safety." Center for Chemical Process Safety of the American Institute of Chemical Engineers. A John Wiley & Sons Publication. Hoboken, New Jersey. 2008.

8-6.   New York State. Department of Civil Service. Workforce and Succession Planning. "Knowledge Management Transfer". http://www.cs.state.ny.us

8-7.   U.S. Chemical Safety and Hazard Investigation Board. "Investigation Report, Vinyl Chloride Monomer Explosion." Report No. 2004-10-I-IL. 2007.

LIST OF FIGURES

List of Tables

INDEX